LETTERS
to
JACKIE

LETTERS

to

JACKIE

Condolences from a Grieving Nation

ELLEN FITZPATRICK

An Imprint of HarperCollinsPublishers

HarperCollins books may be purchased for educational, business, or sales promotional use. For information please write: Special Markets Department, HarperCollins Publishers, 10 East 53rd Street, New York, NY 10022.

A hardcover edition of this book was published in 2010 by Ecco, an imprint of HarperCollins Publishers.

FIRST ECCO PAPERBACK EDITION PUBLISHED 2011.

Designed by Mary Austin Speaker

Library of Congress Cataloging-in-Publication Data has been applied for.

ISBN 978-0-06-196982-9

11 12 13 14 15   OV/RRD   10 9 8 7 6 5 4 3 2 1

For my remarkable mother,

Mary Callahan Fitzpatrick

CONTENTS

INTRODUCTION

"A Deep Scar on Our Hearts"

Please Pardon
this awful Writing
i Cant see very
Well I Feel our
Dear good smiling
Sweet Face our
President is up in
Heaven I always Watch
Him on television
also I Watch his
Funeal. oh Why.
Dear God oh Why.

On November 23, 1963, a day after the assassination of President John F. Kennedy, Katherine Dowd Jackson sat down in her home in rural North Carolina and took out her "letter box"—a cardboard suitcase where she kept white-lined paper and a pen for important occasions. Mrs. Jackson had a third-grade education, but she enjoyed writing. She was moved especially at this moment to express her deeply felt sentiments. "Dear beloved one," she began her letter to Jacqueline Kennedy.

She wanted Mrs. Kennedy to know in the "sad[d]ist moment of your Life you have my great symphy." "I know you are suprized to know," Mrs. Jackson added, "I am a Negro woman." An intensely religious person, Mrs. Jackson had drawn in the past twenty-four hours on her faith to make sense of the President's death. "This marning God spoke to me," she confided. He had told her that the President had done "for his Country what God did for his World[.] They killed our Lord an Father. an now they have killed our Presentend an Father. We loved him but God loved him best."

As Katherine Jackson carefully crafted her message to Mrs. Kennedy, thousands of Americans across the country were writing similar letters. "What can anyone say at a time like this?" asked one correspondent. Few had any answers but many felt an urge to sort out on paper the storm of emotion unleashed by the President's assassination. "As no other First Family has done, you all have come into our homes and touched our personal lives, across the breadth of America. Your voices, your faces, your thoughts, your daily activities . . . were personalized for us," one woman reflected.

Almost a half century later, the events of November 22, 1963, remain a vivid, searing memory for millions of Americans who still recall precisely where they were when they learned of the President's death. Kennedy served as President of the United States for little more than a thousand days. Yet his brief term in office and his shocking assassination deeply touched people of all walks of life, and of every social class, economic station, political sensibility, region, religion, and race. Whether they adored, were indifferent to, or frankly disliked JFK, countless Americans shared the feeling that their own lives would never be the same after their young President died so violently.

The nation has changed profoundly in the decades since President Kennedy's death, as have the lives of all who remember those fateful days. Many of the schoolchildren who raced home on that Friday to discover grieving parents are grandparents today. The "new generation" of World War II veterans that Kennedy's election brought to power has now reached old age. The President's two younger brothers, Senator Robert F. Kennedy—himself a victim of assassination in 1968—and Senator Edward M. Kennedy, are both buried near their brother in Arlington National Cemetery. Wars have been fought. The scourge of legalized segregation has been repudiated. Access to fundamental political and civil rights has widened immeasurably. Fashions and mores of all kinds have changed. And yet for many Americans a filament of recollection easily brings back the incandescence of the early 1960s, when the nation appeared in some ways as bright and as full of promise as its handsome President. Millions still recall how in the passage of a single moment, much of the confidence, energy, and hopeful idealism that Kennedy appeared to exemplify were suddenly swept away. As one young mother predicted shortly after the assassination, "Surely this generation has a deep scar on our hearts which we will carry to our graves."

That scar has inevitably faded in the nearly half century since the assassination. Time has dimmed once vivid memories of the nation's first "television President"—a man whose verve, intelligence, humor, and grace captivated the public. The personal and political mythology burnished in the first years after the assassination have rightly given way in the ensuing decades to a much more complex view of President Kennedy and his administration. Indeed, the pendulum has swung so far in the direction of a more sober, more stark assessment of Kennedy that it is difficult to evoke today the soaring idealism, fresh hope, and sense of possibility that many Americans saw in him. Still, whatever history's judgments about the merits and consequences of those expectations, there can be little doubt that millions of Americans who lived through the Kennedy assassination felt that they had experienced a calamity that they would not forget.

A largely unexamined, and never before published, collection of letters to Jacqueline Kennedy stored in the John F. Kennedy Presidential Library vividly brings to life what the President and his death meant to thousands of Americans in the days and months following the assassination. "How does a nobody write to the wife of our late President?" asked one woman as she began a letter to Mrs. Kennedy four days after the death of JFK. In overcoming her hesitation, this letter writer resembled more than a million and a half other individuals who wrote messages of condolence to the former First Lady. The President died on a Friday afternoon at 12:30 p.m. Central Standard Time. The following Monday, mail delivery to the White House brought a mountain of letters—45,000 on one day—from bereaved citizens, many of whom had sat down within minutes or hours of the assassination to share their grief, shock, and sense of outrage. Piled into cardboard boxes, and then stacked, these containers soon stretched from floor to ceiling, taking up space beyond the offices where "social correspondence" was normally handled and spilling out into the White House corridors. The volume of mail quickly overwhelmed Jacqueline Kennedy's small White House staff, which was nonetheless instructed by the Pentagon to open every single item for "security reasons." On one occasion, loud ticking from a package raised anxieties, until the box was found to contain a wind-up toy sent from Germany to three-year-old John F. Kennedy Jr.

Within seven weeks of the President's death, Jacqueline Kennedy had already received over 800,000 condolence letters. In a population of nearly

190 million, those who took the time to pen a letter to Mrs. Kennedy were clearly exceptional. But the sheer volume of mail, the rapidity with which their messages appeared, the extraordinary diversity of the letter writers, and the parallel manifestations of national grief and mourning evident in the country make the letters a notable element of the public response to the assassination. These individual expressions of grief offer in vivid detail aspects of the widespread reaction to President Kennedy's death.

For millions of Americans, television provided a focal point for the shock, disbelief, grief, and even fears precipitated by the Kennedy assassination. From the moment CBS interrupted its regular television programming at 1:40 Eastern Standard Time on November 22 to report that shots had been fired at the Presidential motorcade in Dallas, the three major networks provided unprecedented news coverage of the assassination's aftermath. For four days they suspended their normal broadcasting and advertising in favor of nonstop coverage of the President's death, lying in state, funeral, and burial. Hungry for stories to fill airtime, the networks ran footage of Kennedy's life and career in an endless loop along with live coverage of breaking events. Nothing comparable in the history of television had ever taken place. And then, on Sunday morning, millions of viewers witnessed in real time Jack Ruby's murder of the President's assassin, as live television covered Lee Harvey Oswald's transfer from one Dallas jail to another. It was, the *New York Times* reported, "the first time in 15 years of television around the globe that a real life homicide had occurred in front of live cameras. . . . The Dallas shooting, easily the most extraordinary moments of TV that a set-owner ever watched, came with such breath-taking suddenness as to beggar description."

Television captured, as well, the crowds that thronged the Capitol. On November 24, some 300,000 lined the streets to watch as a horse-drawn caisson moved the President's casket from the White House to the Capitol Rotunda. For the next eighteen hours, hundreds of thousands filed through the Rotunda—some waiting in line in bitter cold for as long as ten hours. On the Monday of Kennedy's funeral, Tom Wicker would report in the *New York Times*, "a million people stood in the streets to watch Mr. Kennedy's

Nancy Tuckerman (standing nearest the window), *staff, and volunteers sort condolence mail in December 1963.*

last passage. Across the land, millions more—almost the entire population of the country at one time or another—saw the solemn ceremonies on television." In towns and cities across the nation, and indeed around the world, memorial services, eulogies of all kinds, exhibits, and ceremonies remembering the slain American President continued for months afterward.

So too did the outpouring of condolence messages. Mrs. Kennedy's secretaries recruited volunteers to assist in opening, sorting, and acknowledging the correspondence that deluged first the White House and then the Harriman residence in Georgetown where Mrs. Kennedy and her children moved eleven days after the assassination. Some addressed their letters to Hyannis Port where the Kennedys maintained a summer home; the postmaster there estimated on November 30, 1963, that a quarter of a million letters had arrived in the week following the assassination. A few rooms set aside in the Executive Office Building adjacent to the White House soon became the locus of activity, but that space also proved inadequate; additional room was located nearby at the Brookings Institution. The volume of

Dec. 16th 1964

Dear Mrs. Kennedy,

Since this is the first pictures of our girl-boy twins, my husband and I thot it would be nice to send you a picture as the babies were named after you and your late husband.

Jacqueline Lee - 5 lbs. 11 ozs. and

John Fitzgerald - 7 lbs. 14 ozs. were born Feb. 15th 1964. They are getting along just fine now as you can see, although we almost lost John in March.

There is one thing I'd like to ask of you and that is if you'd be so kind to send "two" pictures that the babies may have when they get older. I'm sure they'll be proud of having your names as we are.

Merry Xmas & Happy New Years to you & yours.

Sincerely,

(S/Sgt & Mrs.) William B. Watson Jr.

Stationed at - Fort Benning, Ga.

correspondence ultimately exceeded 1.5 million letters. In May 1965, a House Appropriations Committee report noted that Mrs. Kennedy still received 1,500 to 2,000 letters a week, according to the *New York Times*, "more mail than either former President Harry S. Truman or former President Dwight D. Eisenhower."

Americans also sent artwork, poems, eulogies, Mass cards, newspaper clippings, cartoons, gifts, family Bibles, and military dog tags, among other items, to express their sympathy. Some included snapshots of themselves, their pets, and their children, including the many newborns who had been named after the President or Mrs. Kennedy.

A news report that Caroline Kennedy had broken her wrist prompted a letter to the President's daughter with a picture of the writer's pet dachshund, sporting a cast on his leg. Some sent to Mrs. Kennedy photographs of the President, which they had taken at campaign rallies and other public appearances. Texan John Titmas enclosed in his condolence letter the two poignant photographs of President and Mrs. Kennedy on this book's jacket, which he took at Love Field less than an hour before the assassination.

Jacqueline Lee and John Fitzgerald Watson, twins named after the Kennedys

Mrs. Kennedy is deeply appreciative of your sympathy and grateful for your thoughtfulness

As letters continued to pour in, Jacqueline Kennedy made a brief television appearance seven weeks after the President's death, in which she thanked the American people for their expressions of sympathy. Mrs. Kennedy's remarks on January 14, 1964, gave the public their first glimpse of the

President's widow since the state funeral on November 25. Flanked by her brothers-in-law Attorney General Robert Kennedy and Senator Edward Kennedy, and wearing a simple black wool suit, Jacqueline Kennedy spoke for only two minutes and fifteen seconds. She thanked the nation for the "hundreds of thousands of messages . . . which my children and I have received over the past few weeks." "The knowledge of the affection in which my husband was held by all of you has sustained me," she continued, "and the warmth of these tributes is something I shall never forget. Whenever I can bear to, I read them. All his bright light gone from the world." Noting that "each and every message is to be treasured not only for my children but so that future generations will know how much our country and people of other nations thought of him," she promised the public that "your letters will be placed with his papers" in the Kennedy Library, then already in the planning stages. Her touching remarks, as well as her assurance that each message would be acknowledged, prompted a new avalanche of condolence letters, with many writers apologizing for their tardiness.

In the ensuing months, Nancy Tuckerman and Pam Turnure, Mrs. Kennedy's secretaries, oversaw the monumental project of handling the condolence mail. They supervised a cadre of volunteers who responded not only to each letter but also to requests from the writers for Mass cards or photographs of the President, Mrs. Kennedy, or the former First Family. Most correspondents received in reply a black bordered note card with President Kennedy's coat of arms centered on it, with the simple message: "Mrs. Kennedy is deeply appreciative of your sympathy and grateful for your thoughtfulness."

Sorting the mail itself proved to be a task of gargantuan proportions. A consultant hired to advise about a system for handling the letters admitted

he was flummoxed. Early in the process famed anthropologist Margaret Mead sent word to Nancy Tuckerman, through a cousin of Tuckerman's, that an effort should be made to sort letters into categories, given the value the collection would have for future generations of scholars. Despite the additional workload this entailed, Tuckerman complied. Volunteers separated adult letters from children's, labeled certain items "especially touching," and created various other subsets such as special "requests." Two men from Attorney General Robert Kennedy's staff helped Tuckerman and Turnure create a method of sorting the letters. As late as May 1966, Mrs. Kennedy's staff was still sifting through and cataloging the condolence letters. "Mrs. Kennedy is reluctant to throw things away," Pam Turnure reported to the *New York Times*. "She feels it all came from the heart and who is to know in the future how much any letter or poem or painting will show about how people felt." Of the task involved in handling the condolence mail, Turnure said: "We are doing all we can so that it will be available for work by scholars." And yet, for forty-six years, the letters have sat in the Kennedy Library with little sustained scholarly attention.

When the condolence mail was officially deeded over to the John F. Kennedy Presidential Library in Boston in 1965, it comprised some 1,570 linear feet (that is, the boxes would have extended more than a quarter of a mile if laid end to end). The size of the collection and issues in storage and management posed formidable problems. Eventually these difficulties led the National Archives to pulp all but a representative sample of the condolence mail. The remaining documents represent some 170 linear feet— over 200,000 pages. The team of archivists who sorted through the 830 cartons of general condolence mail saved letters from each major category and created new subsets as they saw fit. All letters offering personal recollections of JFK were retained, as were messages from VIPs, noteworthy mostly for the famous name attached. The enormous volume of foreign mail, some in English but much in the letter writers' own language, was preserved in large measure, organized by country. "Touching," "good," and "representative" letters were also kept—many of them labeled as such by the volunteers who had helped answer the letters. Luckily, the archivists also set aside a random sample of 3 linear feet (approximately 3,000 letters) of the general condolence mail from Americans "as an example of the original inflow of messages to Mrs. Kennedy."

Mrs. Kennedy received very few negative letters. The archivists processing the collection noted an almost "total absence of any letter or comment critical of the president," although several letter writers objected to the President's advocacy of civil rights for African Americans and a very few others viewed him as weak in his attempts to combat the spread of Communism. Presumably those who despised the President found venues other than a condolence letter to express their sentiments. It is in the nature of the genre, of course, that praise, warm memories, and generosity toward the deceased predominate in sympathy messages. The assassination itself, to be sure, contributed to a growing hagiography of JFK that began to take shape from the moment of his death. Still, it is well to remember that during his lifetime Kennedy remained among the most consistently popular of Presidents. Only during the final three months of his life, when a growing number of Americans objected especially to his initiation of civil rights legislation, did his approval rating ever slip below 60 percent—and then only to 56 percent. His average approval rating during his Presidency exceeded 70 percent—higher than that of any other modern President.

The political culture then extant in America surely contributed to Kennedy's idealization. In the early 1960s, the office of the Presidency and the man who held it enjoyed from most Americans a kind of deference that would soon appear a relic of a bygone era. Such fundamental respect and, for some, adulation could flourish, in part, because of a relationship between the President and the press far different from that in contemporary America. An unwritten code of conduct among journalists largely protected John F. Kennedy, as it had previous Presidents, from close scrutiny and publicity about his personal life and, therefore, from scandal. Among the press corps rumors and salacious gossip about Kennedy's private life abounded, but very little of it ever found its way into mainstream news outlets. Kennedy, of course, had his outspoken critics, and the press aggressively covered the ups and downs of the President, his policies, and his administration. JFK never confronted, however, a press likely to expose details of his private life, medical history, or personal relationships that might have undermined the image he projected. Nor did Kennedy face the twenty-four-hour cable news cycle, with its relentless blend of inquiry, opinion, and commentary, much of it giving free voice to highly partisan critics, which would become a reality for later Presidents.

The deep skepticism, with which both press and public often now view government itself, and political office holders of every level, also was far less apparent in Kennedy's era. The infamous "credibility gap" that widened through the long, torturous, and costly engagement in Vietnam, bequeathed to and then inexorably widened by President Lyndon Johnson, badly eroded confidence in the Presidency. Determined and repeated efforts to put the best face on the progress of the American engagement in Southeast Asia, despite mounting casualties and growing public concern that the war was unwinnable, in time damaged the faith many placed in the nation's Chief Executive. That basic trust, buffeted though it often was, reached its nadir in the Watergate scandal during Richard Nixon's term of office. Brilliant and dogged investigative journalism eventually brought down the Nixon White House and with it, at least for a time, public confidence in the Presidency.

The public's sense of personal engagement with President Kennedy was also enhanced by extensive press coverage of the youthful, vivacious First Family. Kennedy's political fortunes soared during the campaign of 1960 in part through his mastery of television. Throughout his time in office, JFK used his facility with this relatively new medium to great advantage. Many letter writers mentioned the utter delight they took in his weekly news conferences—the first to be carried live on radio and television. Kennedy's self-deprecating humor, dry wit, and ability to spar adroitly and easily with the press captivated many viewers. One such citizen recollected, "my husband & I use to get such a kick out of President Kenedy when the News Reporters used to surround him with questions all he had to do is just open his mouth the Answers just flowed out. He never had to study for a minute."

Kennedy's frequent televised appearances clearly entertained some viewers. But he also used the press conferences to explain directly to the

American people his policies and intentions, bypassing the interpretive layer imposed by administration officials and political commentators. "Mr. Kennedy taught my children many things on television," an African American mother from Oakland explained, "because they were interested in him and always wanted to listen to his speeches and my youngest son, Rudolph loved his press conferences and tried to imitate him in many ways." It's been estimated that by the second year of Kennedy's presidency, three out of four adults had seen or heard a Presidential news conference. In 1962, over 90 percent approved of Kennedy's performance. Jacqueline Kennedy, whose sense of fashion, glamour, and interest in the arts enlivened the Kennedy White House, also attracted much public attention. Her February 1962 televised tour of the White House, showcasing her efforts to restore and preserve the historic mansion, drew three out of four television viewers.

With a newborn and a three-year-old, the Kennedys brought the youngest children to the White House of any Presidential family in the twentieth century. The press covered both Caroline and John F. Kennedy Jr.'s activities as extensively as their parents would permit. Many young parents of the World War II generation, who were busy in the early 1960s raising their own children—the baby boomers—strongly identified with the young White House couple.

New media, the culture of celebrity that they enhanced, and the Kennedys' considerable personal appeal converged in the early 1960s to make them the most familiar and closely watched of First Families, all of which enhanced the President's appeal to the American public. Still, JFK's personal qualities surely created some of the magic. Those letter writers who describe even momentary encounters with President Kennedy—a handshake, a wave, a brief conversation—remembered a warm and engaging man whose enjoyment of politics was palpable, who seemed sincere in his convictions, and who appeared to take delight in meeting them. These realities, of course, deepened the personal anguish many Americans felt after Kennedy's assassination. There was much, of course, the public did not know until many years after, but those revelations lay in the distant future.

No historian should construct a biography from condolence letters.

As Jacqueline Kennedy herself noted shortly after his death, Kennedy was a complex man who would, she predicted, always elude complete understanding. His strengths and weaknesses as a President and as a person will likely remain a subject of study and debate for many years to come. But part of what draws people to that debate is the faith and hope so many Americans of such diverse backgrounds placed in the man, and the loss they felt with his terrible, untimely death. In truth, the condolence letters to Jacqueline Kennedy are less about her husband than they are about those whose hearts he captured and their dreams for their country. As one young man wrote to his sister after the assassination: "His death is disquieting to me beyond reason, perhaps, but the death of an ideal is profoundly worse. . . . I am incapable of forgetting his words: 'We must do this not because our laws require it, but BECAUSE IT IS RIGHT.'"

Despite the painstaking care taken to preserve the letters written to Mrs. Kennedy, no academic historian heretofore has systematically read through the entire collection of condolence mail, making it a central focus of sustained study. (The foreign mail constitutes a vast canvas for another historian.) Yet to read this material is to grasp viscerally the enormous impact of Kennedy's assassination on many Americans. In this book, readers will find roughly 250 letters, most of them written by "ordinary" Americans, all of them selected because they help illuminate how some thoughtful citizens, who took the time to record their thoughts, responded to President Kennedy's death. These messages, by their very nature, often begin and end similarly with the writer's wish to offer his or her regret and sympathy.

But through the prism of everyday citizens, the letters included here recount in a very direct and moving way what President Kennedy and his death meant to writers as diverse as the nation. Coal miners, dairy farmers, suburban Republicans, urban blue-collar Democratic Party loyalists, housewives, inmates, schoolchildren, Catholics and Jews, World War II veterans and concentration camp survivors, whites and blacks with powerful responses to Kennedy's civil rights' initiatives, were among the letter writers. Their messages constitute a remarkable record, full of personal anguish and revelation as well as profound meditations on grief, loss, and the human condition. They likewise reveal political passions, prejudices, and perspectives on the nation and the Presidency that predate the chaos and disillusionment ushered in by the Vietnam era and Watergate.

The way in which the correspondence reflects the nation's racial history is alone noteworthy. The early 1960s still belonged to the age of segregation, and there are many letters from African Americans as well as white Southerners that reveal the weight of, as well as the struggle to confront, racial inequality. "We loved your husband because he thought negroes was Gods love and made us like white people and did not make us as dogs," an African American North Carolinian wrote. "I am colored and poor but clean," another woman reassured Mrs. Kennedy in extending an invitation to the former First Lady to visit her at her home in Harrisburg, Pennsylvania.

Most of all, the letters tell the story of Americans united across all their many differences in their shock and abhorrence at an event they could scarcely comprehend. "I am a Florida Dairy Farmer who has been a lifelong Republican. I am Protestant and have been anti-Kennedy since 1960," one man wrote. "However, I feel a desperate urge to extend my deepest sympathy to your children and to you. As an American I am deeply ashamed at the manner in which the President met his end." Many believed they had lived through an event that would alter history's trajectory. Noting that "in two seconds history's course was changed," a young man observed: "The irrationality of life will never be more clearly set down for us. I grieve for John Fitzgerald Kennedy."

My own personal recollection of the Kennedy assassination echoes these sentiments. In the fall of 1963, when I was eleven years old, President Kennedy came to my hometown to dedicate the Robert Frost Library at Amherst College. As every local schoolchild knew, Frost had once lived in Amherst, and it was easy to imagine that he had our small town in mind when he wrote about the New England landscape. One such poem was "The Gift Outright," which Frost had read at President Kennedy's inauguration. "This land was ours before we were the land's," the eighty-six-year-

John F. Kennedy at Amherst College, October 26, 1963

old Frost recited from memory on that bitterly cold January morning that began the administration of the forty-three-year-old President.

Frost died in January of 1963, and Kennedy's visit to the groundbreaking ceremony was meant to honor the poet and repay a favor. I vividly recall waking up with a jolt of anticipation on that sunny October day. As I walked the short distance from our home to the college with my best friend, the smell of wood smoke hung in the air. Before long we heard the President's helicopter whirring overhead. After it made a slow landing on Memorial Field, a motorcade formed, and a pale yellow Lincoln Continental took the President along streets that wound back to the center of the Amherst campus. Kennedy delivered a formal speech inside the "cage," the college's outsize athletic building with its dirt floor and hanging nets, but that event required tickets and was reserved for Amherst alumni, dignitaries, and adults—a group I had not a prayer of joining.

Following the formal ceremonies in the cage, Kennedy was to deliver brief remarks outdoors to the large crowd of town residents who thronged the Amherst campus. The college had planned for a turnout of 5,000 but over 10,000 citizens swarmed the grassy hillside. We pushed our way as close to the front as we could, and then saw a shock of his chestnut hair, heard his distinctive voice, and glimpsed the American President.

Less than one month later, Kennedy was dead. Our sixth-grade class was being given a tour of the school library on November 22 when I heard some staff members saying that there had been a shooting in Dallas and that the President had been wounded. Just that morning my parents had discussed over the newspapers the climate of right-wing extremism that was expected to greet Kennedy during his visit to Texas. I didn't understand the issues, of course, but grasped enough that when I heard the President had been shot, I instantly believed it. I came home after an early dismissal to find my parents staring at the television set as they would until late Monday evening.

My father, especially, was deeply distraught. Born, like Kennedy, in 1917, an Irish Catholic native of Massachusetts, and a naval officer during the Second World War who had served in both the European and Pacific theaters, he much admired Kennedy. On November 22, he told my mother that he felt as if he had lost his own brother. Certainly, the expression of gravity, worry, even devastation on my father's face was one I had never before seen. I understood that something of tremendous moment had occurred. And with this rapid convergence of events—the Presidential visit, the assassination, and my father's evident distress—came for me an acute understanding of how quickly a lived present could recede into the past. These events, without question, shaped my decision to become a historian.

What follows, then, provides an illuminating snapshot of the United States as it existed in 1963 and a rare opportunity to discern how some Americans made sense of a cataclysmic historical event that lingers in our national memory. The political conflict and social ferment that many writers address mirrors persistent tensions in contemporary American society. There was, to be sure, nothing ordinary about the events that inspired Americans to take stock of their own life experience, the fate of their country, and the tragic death of their "beloved President" in November of 1963. "The coffin was very small," as one sixteen-year-old girl observed, "to contain so much of so many Americans." In reflecting on their sense of loss, their fears, and their striving, the authors of these letters wrote an American elegy as poignant and compelling as their shattered and cherished dreams. ❧

November 22, 1963 ·······································

"History Jumping Up Out of History Books"

John F. Kennedy greeting crowds outside his hotel in Fort Worth, Texas, on the morning of November 22, 1963

November 22, 1963, began as any other day in the life of most Americans. Adults went off to work, children to school, the majority inattentive to or completely unaware of the Presidential trip to Dallas. Weddings were being planned, birthday cakes baked, laundry sorted, the weekend and upcoming Thanksgiving holiday anticipated. In hospitals around the country, babies were born, the sick and dying attended to. For some Texans, of course, the day had a different aspect, with thousands planning to catch a glimpse of the President. The Kennedys' visit to Texas was designed to launch the '64 Presidential campaign and smooth over divisions among warring conservative and liberal factions in the state Democratic Party. Kennedy barely carried Texas in the 1960 election, despite Lyndon Johnson's presence on the ticket. He believed his civil rights stance as President might further erode the already wavering loyalties of conservative Southern Democrats. The prospect of a contest with Republican Barry Goldwater made it seem even more essential to shore up declining support in Texas.

The trip began auspiciously. Warm and friendly crowds met the Kennedys upon their arrival on November 21 in San Antonio and later that day in Houston. They stopped to shake hands with crowds at each city's airport, rode in open motorcades through streets lined with well-wishers, and made their way through a busy schedule of events. At every point along the way, eager individuals sought to detain them for a moment. A woman who waited four hours in the lobby of the Rice Hotel later wrote to Mrs. Kennedy, "when you came in President Kennedy shook hands with me." "I'm Mrs. McCockey," she recalled saying. "He looked at me and formed my name with his lips & bowed to me . . . Then I shook hands with you, you probably don't remember but I said Mrs. Kennedy I'm Mrs. McCockey, and you said, how are you Mrs. McCockey." Although the Kennedys arrived at Fort Worth late Thursday night, even then crowds gathered at the airport, along the motorcade route, and at the Hotel Texas where they stayed overnight.

Friday's schedule included a breakfast hosted by the Fort Worth Chamber of Commerce before the short plane ride to Dallas. President Kennedy emerged from his hotel at 8:45 a.m., excited and energized by the thousands who had gathered outside in a parking lot. He apologized for Mrs. Kennedy's absence, noting that she was "organizing herself. It takes her a little longer, but, of course, she looks better than we do when she does it." After offering brief remarks, he waded into the crowd to shake hands and then returned to the Hotel Texas, where he spoke at the breakfast. Some 2,000 people jammed the ballroom and listened as Kennedy extolled Fort Worth's contribution to maintaining national security through its role in military defense construction. Attendees were rewarded with the appearance of Jacqueline Kennedy, beautifully dressed in a pink wool suit and matching pillbox hat. And then they were off for the short plane ride to Love Field in Dallas.

The Dallas leg of Kennedy's trip inspired some anxiety among the President's advisers. Less than a month before, right-wing demonstrators in Dallas had roughed up Adlai Stevenson, disrupting his speech celebrating United Nations Day with boos and jeers and subjecting the UN ambassador to physical violence (as Stevenson left the Memorial Auditorium, two men spat in his face and a woman smacked his head with a picket sign). Leaflets circulating in Dallas the day before Kennedy's visit depicted him as a criminal wanted for treason. Disseminating familiar criticism leveled by the John Birch Society, the handbill accused Kennedy of "turning the sovereignty of the U.S. over to the communist controlled United Nations," offering "support and encouragement to the Communist inspired racial riots" and "consistently" appointing "Anti-Christians to Federal office: Upholds the Supreme Court in its Anti-Christian rulings." It included an old smear alleging Kennedy had a previous marriage and divorce and was lying to the public about it.

On Friday a full-page, black-bordered ad in the *Dallas Morning News* echoed the condemnation. Under a bold face headline reading WELCOME TO DALLAS the broadside accused the President of ignoring the Constitution and promoting aid and comfort to the nation's Communist enemies. After

WANTED

FOR

TREASON

THIS MAN is wanted for treasonous activities against the United States:

1. Betraying the Constitution (which he swore to uphold):
   He is turning the sovereignty of the U. S. over to the communist controlled United Nations.
   He is betraying our friends (Cuba, Katanga, Portugal) and befriending our enemies (Russia, Yugoslavia, Poland).
2. He has been WRONG on innumerable issues affecting the security of the U.S. (United Nations-Berlin wall-Missile removal-Cuba-Wheat deals-Test Ban Treaty, etc.)

3. He has been lax in enforcing Communist Registration laws.
4. He has given support and encouragement to the Communist inspired racial riots.
5. He has illegally invaded a sovereign State with federal troops.
6. He has consistently appointed Anti-Christians to Federal office: Upholds the Supreme Court in its Anti-Christian rulings. Aliens and known Communists abound in Federal offices.
7. He has been caught in fantastic LIES to the American people (including personal ones like his previous marriage and divorce).

being shown the ad on the morning of November 22, Kennedy turned to his wife and commented, "We're heading into nut country today." He went on to reflect, "last night would have been a hell of a night to assassinate a President. ... There was the rain, and the night, and we were all getting jostled. Suppose a man had a pistol in a briefcase." He demonstrated how easy it would have been for an assassin to have fired, "dropped the gun and the briefcase and melted away in the crowd."

Such worries seemed to evaporate as quickly as Fort Worth's early morning clouds and rainy mist. The flight to Dallas from Fort Worth's Carswell Air Force Base lasted just thirteen minutes. By the time the Presidential party arrived at Love Field, brilliant sunshine and temperatures rising into the 80s promised a spectacular day. A Dallas elementary schoolteacher recalled looking up at the sky with her students for the President's plane. "I told the children how wonderful it was that the clouds had lifted, the sun had come out, and you and your husband would have a lovely day after all," she later wrote to Jacqueline Kennedy. "We all thrilled at being so near you—eleven miles—but I felt nearer, for we saw your plane circle in a wide swing before it landed."

Shortly after Air Force One touched down at 11:38 (CST) in Dallas, the Kennedys, who were accompanied by Vice President Johnson and Lady Bird, as well as by Governor John Connally and his wife, disembarked and greeted various local officials selected as a reception committee. Mrs. Kennedy later recollected that she had been given yellow roses at every other stop in Texas, but in Dallas she received a huge bouquet of long stemmed red roses. A few hostile placards were visible at the airport, including one that read YANKEE GO HOME AND TAKE YOUR EQUALS WITH YOU, and another with the blunt, if misspelled message, YOUR A TRAITER. But the overall celebratory mood seemed infectious. A high school student observed that "even though Dallas was mainly a Republican city," the crowds at Love Field were "happy and excited." Within minutes of their arrival, the President and Mrs. Kennedy made their way to the fence line, moving along shaking hands with boisterous and enthusiastic spectators—momentary encounters that would soon be engraved forever in the memories of those they met.

The motorcade left Love Field just before noon. Vantage points for seeing the President were not hard to determine; several Dallas newspapers outlined the motorcade's route prior to the President's visit, and on November 22 the *Dallas Morning News* noted that the motorcade would move slowly so that crowds could "get a good view of President Kennedy and his wife." At the first turn out of the airport, a small group of office workers caught sight of the President. "Just as your car turned from the Love Field entrance onto Mockingbird Lane," one man remembered in a subsequent note to the former First Lady, "Mr. Kennedy was trying to wave to everyone. One girl in our group yelled out 'Welcome to Dallas, Mr. President,' and the President heard her and waved at her. I remarked on the way back to work over and over again that he, Mr. Kennedy, looked beautiful. I know a man isn't usually referred to in this way but this word best described him that day." To this observer, Mrs. Kennedy seemed distant. "You looked as lovely

The motorcade in Dallas, November 22, 1963

as I had imagined you would," he noted, "but just as you passed by us you seemed to be deep in thought and I felt sorry for you as you seemed to be a little weary. I imagined you were tired from the trip."

The motorcade route from Love Field to the Trade Mart, the site of a lunch where the President would give a formal address, covered about ten miles. Once beyond the immediate environs of Love Field, the crowd thinned out for a few miles. Still, when Jacqueline Kennedy put on her sunglasses, the President asked her to take them off, noting that the public would want to see her face. At the President's request, the limousine stopped twice—once in response to a gaggle of schoolchildren holding a sign that said MR. PRESIDENT, PLEASE STOP AND SHAKE OUR HANDS, then to greet some Catholic nuns.

As the motorcade approached downtown Dallas, the crowds swelled, excitement built, and cheers rang out in places where spectators stood as many as twelve deep on the sidewalks. Flags fluttered, office workers leaned out of windows in tall buildings, and some intrepid people stood on the awnings and roofs to get a better look. Dallas policemen struggled in places to hold back the surging crowd. As he waved, the President murmured,

"Thank you, thank you" again and again. Traveling at a speed estimated between seven and eleven miles per hour, the President's car allowed some spectators memorable impressions of the Kennedys. One Catholic nun reported to her parents, "We were so close to them that if I wanted to, I could have reached out and touched the car." "He looked so darling and he had a real wide smile and his eyes were real bright," she remembered. Mrs. Kennedy offered a "big smile and her graceful wave" and then the President himself "caught sight of us and turned toward us and waved and said 'Oh the Sisters.' Then it was over all too soon." Down the twelve blocks of Main Street the latter sentiment arose again and again. The Kennedys were there for a moment—smiling, vibrant, alive—and "then they were gone."

The rapidity with which events next unfolded remains one of the more stunning facets of the Kennedy assassination. The motorcade came under fire at 12:30 p.m., just after it had zigzagged from Main Street to Houston and then around the corner to Elm where the Texas Book Depository stood. The crowds thinned past the Depository. Jacqueline Kennedy waved and looked to her left, avoiding having to gaze directly into the sun. As the heat of the day beat down, she anticipated the relief that would come when they reached the cool underpass ahead. Many bystanders heard the crack of the rifle as the first shot rang out. Another followed in rapid succession. Mrs. Kennedy at first imagined the sound was a motorcycle backfiring until Governor Connally, sitting in the jumpseat ahead, cried out. She saw Connally grimacing and hitting his fist against his chest, and then turned toward her husband. He had a "quizzical" expression on his face, she recalled, as he raised his hand as if to smooth back a lock of his hair. She leaned toward him, now only six inches away, when another crack of the rifle pierced the air. The third shot delivered a lethal wound to the President's head, showering Mrs. Kennedy, a motorcyclist nearby, and the limousine with gore. Kennedy slumped toward his wife, the backseat now "full of blood and red roses," she would later recall. Two blossoms, lodged inside the President's shirt, would be given back to her later that night when his body was returned to Bethesda Naval Hospital in Washington.

Spectators reacted immediately to the sound of gunfire, many running away from the street or throwing themselves to the ground. Several of those closest to the Presidential car turned toward the Book Depository when they heard gunfire. Among them was Bob Jackson, a photographer for the *Dallas Times Herald*, who was seated in an open car reserved for cameramen toward the rear of the motorcade. He looked up in time to see a rifle being pulled back from a sixth-floor window in the Depository. Tom Dillard of the *Dallas Morning News*, also in this convertible, quickly snapped a photograph of the sniper's perch.

Before the Presidential limousine even reached Parkland Hospital, a distance of less than four miles that nonetheless felt like an "eternity" to Jacqueline Kennedy, news of the assassination attempt began to break. Merriman Smith, the White House Correspondent for UPI, was riding in the press pool car just in front of the photographers when he heard the gunfire. He grabbed the radiophone and called his Dallas bureau, shouting: "Three shots were fired at President Kennedy's motorcade in downtown Dallas." The bulletin came off the UPI teletype at 12:34, setting in motion a cascade of breaking news stories. At 12:36—just as the President's car reached Parkland Hospital—ABC radio interrupted its programming to read the UPI flash. At 12:40 CBS broke into its popular soap opera *As the World Turns* with a special bulletin read by Walter Cronkite: "In Dallas, Texas, three shots were fired at President Kennedy's motorcade in downtown Dallas. The first reports say that President Kennedy has been seriously wounded by this shooting." Updates by wire report, radio, and television rapidly tumbled in. By 1:00 p.m., when doctors at Parkland pronounced the President dead, it's been estimated that nearly 70 percent of adults in the United States already knew of the assassination attempt. At 1:35 another UPI bulletin came across the wire: "President Kennedy dead."

As these events unfolded, word of them reached Americans who were going about their day only to be stopped short. Their subsequent letters of sympathy to Mrs. Kennedy reveal how the drama of November 22 collided with the lives of individual Americans, both on the scene in Dallas and far removed from Texas. Some began their letters as they sat watching news of the assassination break on television. There are no letters from eyewitnesses in the condolence letters, but there are many from bystanders who saw the President and Mrs. Kennedy only minutes before the assassination. Others wrote weeks, months, and even a year later, but described with extraordinary clarity precisely what happened in their own lives on November 22. The first letters below are arranged chronologically, based not upon the date of their letter, but of the day's events. They are followed by messages that depict the way the news reverberated around the country. ᷾

Dear Mrs. Kennedy,

Nothing has been confirmed as yet. Either way it turns out you have my deepest sympathy.

Don't neglect the healing affect of quiet time alone with a quiet horse or dog.

Oh my dear, it has been pretty well confirmed that we have lost him. Our prayers are with you.

> Love,
> Nancy, Kenneth,
> Rick, Brandon
> & Perry Glimpse

UPPER DARBY, PA.
11/22/63
2:00 P.M.

My dear Mrs. Kennedy.

Even as I write this letter, my hand, my body is trembling at the terrible incident of this afternoon. I am watching the CBS-TV news report. No official word as yet. I'm not of voting age yet but I am old enough to understand the political and diplomatic relations of the world. When the President was campaigning, If I had been old enough, I would have voted in his favor. I knew that then and I know that now. Not because of his youth, his religion, his personality. But because of some indistinguishable influence, perhaps more defined now after a few years, almost a full term, after his election.

I'm writing, I know, but what I want to say, I can't put into words. Perhaps you can read between the lines. Not just "I'm sorry to hear. . . ." but more.

Good God! Help us! Help us! Three assassinations are now history and I never thought I'd live to see one, even a thwarted attempt which was close but not, Saints help us SUCCESSFUL. It is a terrible thing to live through.

I can't go on writing now. It's too much. My prayers are with you and those involved.

Larry Toomey
11/22/63 2:45 P.M.

DALLAS TEXAS
DEC. 1 – 1963

Mrs Jacqueline Kennedy

First Lady in our hearts.

I live in Dallas, a city bowed in sorrow, and shame. I am 76 years old and live on a social security check

I must pour out my heart to you if my feeble hands will hold out to scribble a few lines.

I was at Lovefield, when you and John steped from the plane. I was the first man to shake his hand, (from behind the fence barricade). That was my life's fullest moment.

And you! The camera's were on you, most of that dark day. Heaven must have fortufied you for those hours. No Pen or brush nor gifted tongue could have accurately portrayed your stature in lonliness. Humility, bravery, fortitude, beauty, strength, faithfulness, loyalty, loveliness, and grandure, Rolled into one Sweet Mrs. Amerec of the Ages.

Very Sincerely,
J.E.Y. Russell

Dear Mrs. Kennedy,

The following letter is a copy of a letter that I wrote to a parish priest (Episcopal) who moved away from this area some time ago.

I thought perhaps you would know by this letter that there are those who will never forget your husband and who will always miss him. Even now, six months afterward, unexpected tears spring to my eyes every time I see a film of him on television. Even now it is so hard to believe. I whisper to myself, "Surely this can't be so!"

Your beautiful picture on the cover of <u>Life</u> and your article prompted me to write to you. I hope I have given you some comfort.

Most sincerely,

Janice Crabtree

(Mrs. W.C.)

November 27, 1963

Dear Father,

May I share a few thoughts with you about the tragedy? Nothing has touched me so deeply in a long time. I had seen President Kennedy just three or four minutes before he was shot. I had planned all week to go to the parade in downtown Dallas, but the morning dawned foggy, misty and ugly. Billy insisted that I stay home and watch the motorcade on television. But by 9:30 a.m. I couldn't sit still any longer. I put on my oldest raincoat and overshoes and dashed to Dallas. I parked way down on Pacific, and was the last car that that lot could take. Excitement was in the air, and I was glad to be alone so I could soak it up without the necessity of polite conversation with anyone. I walked slowly, trying to kill the long wait. . . . Finally I decided to go to Neiman's Zodiac Room for a snack, but they were having a private brunch until noon, so I sadly turned away. As I did so, I saw the Beauty Salon, and right there decided to get my hair cut. I was happily surprised that they could take me. I told them that I couldn't wait, because I wanted to see the President.

When I came out of Neiman's with my new haircut at 11:10, crowds were already forming. It was quite heartening, because I had worried so about his reception in Dallas. I hurried down to Daddy's old office building, the former Republic Bank Building, now the Davis Building. Jacqueline Kennedy is not the only one with outstanding sentiment—I wanted to see the President right in front of the doorway that my father used thousands of times, and I wanted to try to imagine and feel the elation that he would have felt about seeing his favorite of all presidents. To Daddy, President Kennedy was too good to be true—he worried constantly about an untimely death for him. I remember his astounding statement at the time of the 1960 Democratic Convention, when Johnson accepted the vice-presidency and so many of Johnson's fans were sick— Daddy said, "Kennedy will be elected, then assassinated, and Johnson will be president, after all." I thought of this during the long wait. I looked up at the windows of the tall buildings and thought about the utter futility of trying to protect him. The thought crossed my mind that a bomb tossed from one of those windows could kill a bunch of us, too, but even that did not induce me to move from my perfect spot. The crowd grew and grew. Soon rooftops and awnings were crowded. Right across the street an enormous sign was put up. It said, "We'll trade U one retired general for some NASA artwork. Signed 250 Dallas artists." Police cars made constant patrols, looking, watching. A police truck hauled off a car that was left on Main Street.

I wondered what the owner of the car would think when he returned and couldn't find his car. The crowd was very jovial and those of us who shared the long foot-tiring wait became like neighbors.

A couple of incidents brought real laughter from both sides of the street. One was a powder blue car with bright writings all over it. It would go past us, catching all eyes, then pretty soon it would come back. We began to clap on about the third pass, and on the last we cheered—the writing on the car said, "Shop at Honest John's Pawn Shop," and the shrewd owner was taking advantage of the great crowd. The other incident had to do with two high school boys in a convertible—one was driving and the other sat in the back seat smiling and waving with an expression on his face like a great leader. The amiable crowd rewarded the boy with light applause and good humor. I heard only one ugly remark about President Kennedy—a squat sour-looking old man came out of Daddy's doorway, pushed his way to the curb, looked at the size of the crowd, and said, "All you people here to see that guy?" Of course, no one answered him and he hurried back inside. A young girl next to me had a transistor radio, and we were able to hear on-the-spot reporting about his wonderful welcome at Love Field, about his friendly handshaking, Jackie's beauty and everything—the excitement was mounting. Finally, the police turned away all traffic, and Main Street was empty at noon. The police cautioned us to stay on the curb, but we couldn't resist dashing out into the quiet street for a long look to see if the motorcade was approaching. At last it came into view, and that first sight of it filled me with such incredible excitement that I don't believe I can describe it—indeed, even to write of it starts my heart pounding. The first thing I was able to see at several blocks distance were the red lights of the motorcycle police escort—about eight flashing red lights preceding the dark limosine. They were travelling faster than I had expected. The police were yelling to stay back, but from both sides of the street we surged out. I almost got my toe run over by one of the motorcycles. Long as I live I will never forget Kennedy— tanned (that was the first thing that I noticed) smiling, handsome, happy. I didn't get to see Jackie's face, because she was waving to her side of the street, but her youthful image was unmistakably beautiful. Her long mahogany-colored hair was blowing in the wind, and the sun, which had come out brilliantly, caught the red highlights. I'll always remember the way it shone so brightly on the President. Then they were gone.

I was shaking so, as I made my way back to the parking lot, that I decided I'd better stop on the way home and eat a bite of lunch. I got into my little Opel and turned on the radio. The first thing I heard was, "The President has been shot," and I just thought that the announcer had meant to say that the President has been <u>shocked</u> at the size and friendliness of the crowd. All too soon the terrible truth sank in and I don't know how I got home. I couldn't go into my house alone, so I went to the neighbors. They were white-faced and weeping. Their television was on and the announcer had just said that our good, wonderful, youthful President was dead. One of my neighbors had attended the breakfast in Fort Worth that morning—

Needless to say, I never did eat lunch or supper. I never made up a bed, got together a meal, nor paid a bill until after the funeral. I even almost forgot that my beloved Mother had died on November 24th, and was buried on Billy's birthday,

November 26th. Today is my own birthday and it means
nothing. I am sick, sick, and violently angry.

Pray for us all, please.

Janice

P.S. Would you mind sending this back to me? I would like to keep this
written account for my sons, Billy and Jim.

Dear Mrs. Kennedy:

I know the grief you bear. I bear that same grief. I am a Dallasite.
I saw you yesterday. I hope to see you again. I saw Mr. Kennedy yesterday. I'll never see him again. I'm very disturbed because I saw him
a mere 2 minutes before that fatal shot was fired. I couldn't believe it
when I heard it over the radio 5 minutes later. I felt like I was in a daze.
To Dallas, time has halted. Everyone is shocked and disturbed. My
prayers to you.

A Sympathetic, Prayerful, and Disturbed
DALLASITE,
Tommy Smith
Age: 14

---

*S*ome bystanders waiting to catch a glimpse of President Kennedy instead saw
the motorcade as it sped toward Parkland Hospital. One woman who later
wrote to Mrs. Kennedy received a call from her brother who witnessed the arrival
of the presidential limousine at the hospital. "Calls for stretchers rang out—and
you know too clearly the rest," she noted. "John, my brother, helped take your
John, by stretcher to the Emergency Room. . . . I saw my brother around 2:00
p.m. that afternoon. He was visibly shaken." A nursing home administrator who

happened to be at Parkland that day with a patient recalled, "I saw your beloved husband when they brought him in on the stretcher to the emergency room. . . . I am sorry I was unfortunate to have to see your wonderful husband on his day of death." Others waited for the motorcade to arrive at the Trade Mart.

DALLAS, TEXAS

Dear Mrs Kennedy,

I would like too express my sympathy in losing your most wonderful husband

Mrs. Kennedy I was one of the ladies choosen from my Church St. Pius X to serve the dinner at the Trade Mart.

I was so proud and thrill too know I would get to see you and your beloved husband and all the party with you all.

I couldn't hardly sleep or eat knowing that we ladies were going to serve.

We ladies had everything prepare and just waiting for you all too come too the Trade Mart. My heart just fell when we heard the sad news and I felt like it was one of my family. I wish I could do something more for you and your children. But the only thing I can offer is too pray to God for you and your children.

I will never forget you and Mr. Kennedy. I pray for his soul every night and ask God too take good care of you and your children. May God Alway Bless You.

Mr & Mrs. Frank Cuchia

JANUARY 6, 1964
DENTON, TEXAS

Dear Mrs. Kennedy,

Regardless of my knowledge that you will probably never see this letter, I am writing to express my deepest sorrow over the death of your husband.

Having grown up in a strongly Republican family, I was most disheartened after the 1960 election. I was not a Kennedy supporter; in fact, I was bitterly disappointed when he won. It has been only in the past few months that I had come to a realization of your husband's greatness and value to us all. On the eve of your Texas visit, I finally reached the conclusion that President Kennedy was a great president and that I would vote for him in the 1964 election.

I tell you this boring history that you may know that my overwhelming and lingering sorrow at his death came not from an emotional attachment to an extremely popular man and name, but as the result of serious consideration in light of his proven ability.

My anguish at his death was probably heightened by the fact that on the morning of Nov. 22, I waited in front of the Trade Mart in Dallas to catch a glimpse of you and the President. To this hour I can still feel the same horror and grief I felt then when I understood the delay in reaching the place where I stood.

As a student of history, I know the memory of the American public to be short. In the coming months many people will forget the horror of that black Friday and the exhilaration of your husband's administration. But I will not forget. And I join with you in the belief that "there was once a spot; for one brief, shining moment that was known as Camelot."

Most sincerely,
Carol Oakey

DEC-17-1963
TEAGUE, TEXAS

Dear Mrs. Kennedy, Carolyn and John John,

I am a person of poor education as you will see, but I wrote President Kennedy a letter of gratitude for all the many things he had done for us the American people and as I was sending out greetings to friends and loved ones, I felt impressed to write our Dear President, so in my feeble way I sent a greetings to him and you his wonderful family last year. And I received a letter of thanks from his secetary to my happy surprise.

I am still stuned and grieved over the horrable death he met so suddenly on Nov – 22 – 1963 in Dallas Texas. I had watched on T V your visit to Sanantone, Houston, Ft worth and Dallas, which is 100 miles to the north of us. I was watching the parade and thinking how Happy you and our President were over the warm welcome from the people of Dallas, when at a split second we noticed that the presidents car also the Vice presidents car, had turned and left the parade, of course we were very upset, then the news flash just in seconds gave us the sad news that our Dear President had been shot.

My daughter and I were having lunch celebrating her birthday the 22nd we stoped and folding our hands in prayer we ask the Good Lord, if it was his will to be with our president to heal, and spare his life, but in just a short time there was another news flash, that our wonderful President had died from the cruel snippers bullet. And I want to commend you Dear mrs Kennedy for your great courage at that time. And altho I'm sure your life and the lives of your darling children can never be the same, I believe in God and I will continue to pray for the presence of the Lord to ever be near you and that his holy angles will watch over you and your children and that you will have the peace of God in your heat that passeth all understanding during this Xmas season and on through life.

Mr and Mrs J Harper

CHARLOTTE, N.C.
JANUARY 17, 1964

Dear Mrs Kennedy,

May I extend my heartfelt sympathy and prayers to you and your family? I do so admire your courage and strength. It has been an inspiration to me the way you have conducted yourself. You see Mrs Kennedy, my husband died of a heart attack while sitting at the table drinking a glass of milk on Friday, November 22, at about the same time your husband and our beloved President was killed. We were listening to the news about your husband when my husband had his attack. His last words were "how could anyone have such hate in his heart that he could do such a thing to our President." My sixteen year old son came in at that time from school and was with me when he died. He died not knowing for sure the President was dead. My husband was 46 years old also—born in April, 1917. We have five Children—four boys and one girl. Our oldest boy is 22 years old and doing graduate work at Brown University in Prov. R.I. My youngest is nine years old and in the 4th grade. My husband was a retired Army Major.

I feel so for your young children. Mine are older and will remember their father real well.

My prayers will be with you and your family in the difficult days ahead. I can truly sympathize with you as I am going through the same adjustment—that of adjusting your life without the man you love by your side.

Sincerely, Margaret McLean

Millions of American children learned about the Kennedy assassination in school. As they sat in their classrooms on Friday afternoon, word of the shooting in Dallas began to reach school officials and teachers. Many of the latter confronted the unenviable task of breaking the news to their pupils. These educators described the anguish that task imposed as they struggled to maintain their own composure. The young children they taught looked up at them with open faces searching for guidance. Often unsure how to respond, teachers improvised assignments and activities they hoped would reassure and occupy their stunned pupils. Few of those children, however, missed the impact of the assassination on their mentors. In their letters to Mrs. Kennedy, they wrote with simple eloquence about how upset their teachers were and how difficult it was to make sense of the news, however it was delivered. They also often commented on the distraught parents they found when they arrived home from school on November 22.

NOV. 22 1963
FORT WORTH, TEX.

Dear Mrs. Kennedy,

I was at school when I heard about the President. I cried for two or
three minuts. My mother also cried, and so did my teacher Mrs. Mansir.
I was very sad for President Kennedy. He was my friend even though he
didnt know me. Some of my school mates hid their faces in their arms in
sadness. I told my father that I wish we could have you and your children
to care for. Today I saw you and your husband at Carswell Air Forse Base.
He was happy. It was terrible to have him shot. Ive been watching T.V.
sense 3-40, friday. I respected him, I liked him. Would you please if you
can send two photographs of you and President Kennedy. Thank you.

David Blair McClain

···❖

BEACON, NEW YORK
NOVEMBER 22, 1963

My dear Mrs Kennedy,

I know that in a tragic day like today, I shouldn't be silly enough to
write you a letter. And I know you probably won't read this, and I will
never have an answer. I am writing this in school. During the changing
of classes there was a rumor that your husband had been shot. Nobody
believed it. Nobody wanted to. I went back to my room and told my
teacher about this rumor. She didn't say much. About 5 minutes later,
the principal announced over the intercom that the President had passed
away. My teacher broke down in tears. No one else did because no one
believed it. We got a radio and sat in class to listen to the news. Slowly, as
I looked around, I saw everyone break down. The principal, the superin-
tendent of schools and everyone was crying. The nurse came to my room
to say that her office was packed with hysterical pupils, and that if any
of us wanted to go to her office we could. I went because I was a nervous

wreck. As I walked through the halls I noticed an unusual calm. I knew the rumor was true. I have grown up with[out] a father. Last year the head man on my list, my grandfather, passed away. Since then your husband was the man I looked up to. Your husband was the greatest President that I have ever heard of. I feel that I knew him as a man and a friend, not the head of a country. As I write this letter I burst into tears, over the loss of a great man. You had a great husband, and his memory will last forever.

Thank you for listening.

Sincerely,

Nancy Ashburn

P.S. I wish they would let me get my hands on the assassinator.

11/22/63
LYNBROOK, NY

My dear Mrs. Kennedy,

Now that I have started this letter I find how difficult it is to express sympathy for a loss as great as yours. Some would resort to tears. Some may be shocked into silence. Some would merely shrug and try to pretend that it never happened. My school showed sympathy with a traditional "half-mast" of the flag.

I cannot be traditional nor shrugging nor silent nor tearful. Many times I keep my deepest feelings to myself as I am keeping this expression of grief a secret from my family.

The news of Mr. Kennedy's death was announced to me between classes. I felt, of course, astonishment then pain. A knot built up in my throat and tears threatened to overflow. My mind was briming with denials of the cold, hard fact that the President was dead. After school I hurried home only to find my news had reached there before me. On the way home I glanced at people going about their every day lives and I wondered if they knew. I was the messenger, the Mercury, to tell all of them what had happened. I tried to yell but the words stuck in my throat and found no release.

The only relief I felt was upon stroking my dog's head and watching the blue violet clouds go by, seeming to know of the tragedy for all their color.

Your husband's death reminded me of another in Stephen Crane's The Red Badge of Courage:

> "... I see' a feller git hit plum in th' head when my reg'ment was a-standin' at ease onct. An' everybody yelled out to 'im: Hurt, John? Are yeh hurt much? 'No,' ses he. He looked kinder surprised, and he went on tellin' 'em how he felt. He sed he didn't feel nothin'. But, by dad, th' first thing that feller knowed he was dead. Yes, he was dead—stone dead....

The reference may be harsh but Mr. Kennedy was in a battle too. He was a soldier in the regiment for peace. He was known throughout the world.

I implore you Mrs. Kennedy to pray, not for revenge, but for strength to continue. I am not a prominent head of state. My message will probably not reach anyone's eyes but yours. I only want you to know that it carries the grief of a nation and of a fourteen year old, me.

I feel for you and sympathize with your grief.

Sincerely,
Susan DiGeorgio

....✤

FORT WORTH, TEXAS

Dear Mrs. Kennedy,

I am a Catholic also, I go to Saint Georges School. I can remember Nov. 21, the day before you came. We go to mass every day, then we go to lunch. This day was different, after mass our pastor told us to sit down. I wondered to myself "Whats going on?") Then he told us some wonderful news "Our wonderful president and his lovely wife are comming tomorrow." Then terrible news "Only the 7th and 8th graders are going," and wouldn't you know I'm in the 6th grade! I couldn't stand it! They got to go and I dident!

But we got to watch you on T.V. (There was such a big crowd, they didn't get to see you anyway.) When you dident come, we got worried. Our class clapped when President Kennedy came, yet when you dident arrive, it left us wondering. We noticed that he didn't smile. His first realy and truly grin, was when he saw you. Everyone in our classroom cheered and appualed. What excitement.

After mass we went to the chior pratice. I am an alto. We heard the phone ring, and after a moment she hung up (our principle Sister Leanardice) and rushed away. Everyone but me, was calm. I sencted something wrong. Then the mikeraphone was turned on, an a sad voice said "Kneel down and pray, the president has been shot.

Nearly everyone broke into tears. I tried to control myself. Then we heard that he was dead. I couldn't believe it. Nancy Keeney who can't controle herself, and who had cried sence he had been shot fainted, I almost did. We went for the lowering of the flag. Then we said a rosary for him. Everyone even me, cried, except for Debra De Milo, all she worried about was war. I got mad and told her "Fine time to worry about war. The president is dead, worry about him, he's more important.

That evening large headlines covered our newspapers front page. Kennedy slain, Connally wounded. I wish I was as brave as you.

My sister goes to Incarnet Word said that his hair was surprisingly light.

If at all possible send me a picture of your family. Also if you have any time tell me a little about your family.

Mary McMillen
At least I can say, "The president had his last meal in my town. I can also say "I was born in Warshington D.C. (Doctors Hosbitle)
Yours Sincerly
Mary

NOV, 23, 1963

Dear Mrs. Kennedy,

As I and many other people of the United States heard the shocking
new of our late husband Mr. J.F. Kennedy made us bust out in tears. When
I heard the shocking news the tears came out of my eyes very fast. I felt
so bad. Then all of a soden I stop crying and thought about you. As I was
thinking I said to my self I said how does poor Mrs Kenedy feal as she sat
beside her husband when he was shot. Then when I went home from school
I found my grammother sitting down in front of the television set. And I
could see she was crying. So I said what's the matter gram. She said she had
heard the shocking news, and was thinking how you must of felt. Because in
1926 she was setting on the front pourch of her house with my gramfather
when a group of man drove by the house in a car and shot my gramfather in
the chest 3 times. Instintly he died. So she said she nows how you must feal.

I send my greatest simpathy to you and your family.

> Very Truly yours
> Jo-Ann Palumbo

P.S.
I think God said he wanted your husband the man who searved his country well to join him in heaven now and no other time. God bless you and
your children.

> Jo-Ann Palumbo

VASHON ISLAND
VASHON, WASHINGTON
NOVEMBER 24, 1963

Dear Mrs Kennedy,

Because of Caroline's age, I thought you might appreciate knowing how my first grade class at Gatewood Elementary School, Seattle,
Washington, reacted to the news of Friday's tragic event.

I was seated with a group of "Dick and Janers" when a typed bulletin came to me from our school office. I glanced at it, expecting the usual rainy-day recess or some such announcement. Totally unprepared for its content, I gasped audibly and sat in stunned silence forgetting that the wide eyes of twenty-eight six-year-olds were upon me. One little girl brought me to with the question, "What's the matter, Mrs. Mackey?"

The impulse came to spare them the news, and then I felt I had no right, young as they were, to rob them of living history. I said simply, "President Kennedy has been shot." Familiar with the "good guys" of Westerns, one little boy said, "But he can get well." Just then, our secretary came in to inform me that the anguish in my heart would have to stay. I told the children of the president's death.

We have a television in our classroom and I switched it on. A priest was praying and we, the children and I, stood in silence. The coincidence that caused me to break the ruling of the Supreme Court outlawing prayer in a public school surely cannot be held against me.

I turned off the television and listened to the children's chatter too shocked to move for a time. Some of their expressions stay with me: "That's Caroline's Daddy and I feel awful for her." "I liked President Kennedy because he was so good." "I'm going to say prayers, that's what I'm going to do." The rapidity with which these little ones grasped this terrible tragedy and their warm and spontaneous expressions of sympathy were amazing.

Soon the school flag, visible from our windows, was lowered to half-mast. We discussed this symbol of mourning and then they wanted to write about it. Large pencils in twenty-eight little hands wrote the history they were living:

"President Kennedy is dead. Our flag is at half-mast."

They all seemed to want to print their best. Then with crayons they drew the flag at half-mast. Old Glory had some bizarre stripes, but each little artist went home with the dawn of patriotism and has sorrow for a fallen hero.

As I am writing these lines, downstairs in our home, my husband is seated at an ancient organ surrounded by three parishioners of our small island church, St. Patrick's. Their voices and the organ are earnestly pouring out the Requiem which will be sung tomorrow at 9:30 A.M. for our beloved president.

Many miles away in the Jesuit House of Studies, Springhill College, Mobile , Alabama we are sure our only son, a seminarian, will be singing such a Mass too.

With deepest sympathy,
Vivian Mackey

WHITESTOWN, IND
JAN 20 1964

Dear Mrs. Kennedy,

I am sorry that your husband died. When I heard about it I was in school. When I came in my teacher was crying. He wrote on the board what happend, when I saw that my heart skipped a couple beats. Then all of the sudden I felt sick. We watched everything on T.V. and went to church and prayed for him.

I sure felt bad about it. I am interested in him so I gave four reports on him.

After we got back in school this one girl asked why you couldn't be President. The boys said because women weren't smart enough. But I said if it weren't for women men wouldn't be here so that was the end of that. Bye for now.

Patricia Anne Hemmerle

Dear Mrs. Kennedy:

This is not the first letter I have started to you, for twice before in the weeks that have passed I have attempted to write, but could not. I think that the numbness and shock of personal grief which has been felt by the whole nation and the rest of the world had its grip on my own heart in such a way that words could not be found to express it.

I am not sure that now it is possible. Perhaps in giving you just a bit of my own life I may say it more clearly.

I am a teacher of second grade children. Our little town of La Porte here on the Gulf coast is within commuting distance of Houston and two of my pupils at school were with their parents among the thousands who lined the streets of Houston the day you were there. It had been our pleasure to have the newspaper picture of John John (in the open door beneath his father's desk) on our bulletin board for several days. It had been brought to school and put there by one of the children.

So aside from the fact that they had actually <u>seen</u> <u>their</u> <u>president,</u> the report of the day in Houston had special meaning for the other children because of John John (I think you should know that little children <u>love</u> that picture and that it is one of our most prized possessions).

The TV program at school brought us this tragic news and it was as if it had been a member of their own families.

You can not know these things unless we tell you, and I am sure you have had many thousands of letters come to you. Somehow in writing my own I have felt that perhaps I might let you know how <u>children</u> feel, too—for it is a <u>personal</u> sorrow with them, as with adults the world over.

It has, indeed, been a personal one with my husband and me and with our friends and fellow teachers. I can only <u>tell</u> you of this—how we met in our principal's office and listened, and cried together as a faculty, and how we sent our students home that day with the knowledge of national tragedy and sorrow a part of that day's education. But I hope and pray that it may be a source of comfort to you now to know that each day will bring its own opportunity to all of us to strengthen and carry on in our education of even the very young children the direction which your loved one gave his life for. His ideals live on—just as he will—in the hearts and minds of all of us.

We have an only son who is now a graduate student in Duke University. It was his birthday and when he was at Texas University in the four years just past, we have always phoned on the birthdays he could not spend at home. That night we called him in North Carolina. He is a member of a group of twelve graduate students, admitted in June for the two-year program in Hospital Administration at Duke. He shares quarters with another member of the group, and usually we get the roommate when we call, but this time our son, Stan, answered.

It was his birthday (his 24th) and of course if it were to do over, I would still call as we have always done in the past. But when he heard my voice he said "Oh Mother—" and began to sob. I have not heard him cry since he was ten, but over 1500 miles of distance throughout my call he could not talk—he only cried.

Young men in graduate school, parents at home, little children who saw their president in Houston, a principal and his staff of 22 teachers—this is a small cross-section of our great nation, but it happened to be the part that I had knowledge of at the time. It is for these people, as well as myself, that I write to you now. It may be some time before you read my letter in the many many others which I am sure you have yet to give attention to, but I do hope that some day this one may reach you, to bring you our <u>love</u>, and our earnest prayer that you will always remember he is not lost to you—nor to us. And what he meant to our Nation will live on. It will live on—through graduate students, their teachers, and through little children who will learn from their teachers, and parents and others, the concepts of greatness which were so much a part of your dear husband.

We were privileged to see it in <u>you</u> in the days that followed—<u>true courage</u> and <u>dignity</u>—all that American womanhood could ever aspire to! God bless you and keep you, and may His special gifts of tenderness and love be with you and your precious children, in all the days to come.

Sincerely,

Irene Lowrey

(Mrs. G. C. Lowrey)

DECEMBER 1963
PONTIAC, MICHIGAN

Dear Mrs Kennedy,

I was shocked to hear of your husband's death.

I was coming home from school and was feeling fine. My mother had tears in her eyes when I saw her. I asked her what was the matter because she had tears in her eyes They didn't tell use about the tragedy at school. My first though was that is wasn't true. I wish it wasn't. But I turned to her and her eyes had truth in them. I broke down and cryed.

It was like a nightmare for the whole nation. The world died a little bit it self when John Fitzgerald Kennedy died.

This is something I will never forget. I am 11 years old. I wrote Mr. Kennedy a letter after the eletion to tell him how happy I was he had won.

and when I was looking at the funeral on telivison I cryed through the whole thing It was so sad.

I still have the letter that he sent me. I will show it to my children when I grow up.

President Kennedy will never be forgotten in the United States of America.

Yours truly,

Nancy Taylor

Dear Mrs. Kennedy,

As I sat watching the T.V. set this afternoon, I decided to write to you and to extend my sincere sorrow and that my fellow-eighth graders at St. Clare's School, California.

On the morning of November 22, our school of 750 pupils were at a requiem Mass for all the deceased of parish. At the beginning of the Mass, we were told that our beloved president was shot. I tried to tell myself he would be all right but somehow I knew he wouldn't. I tried to control myself as I had to play the church organ but the tears wouldn't stop. The slightly damp keys were hard to play but I offered it up that the President might live.

Though we didn't know it then but while 750 children with tear-streaked faces and slightly reddened eyes were receiving Holy Communion, the 35th President of the United States went to his eternal reward in heaven. I firmly believe that your husband is sitting up in heaven— right next to Lincoln.

Though I never knew President Kennedy or so much as saw him except on T.V. and in pictures, I feel as though I have had the pleasure of meeting him in person. If I live until 103, his memory will live on within me as I'm sure it will within all his personal friends and especially you, his beloved wife.

Please accept my sincere sympathy and my many prayers.

May God give you courage for the years ahead and bestow upon you many blessings.

Sincerely

Mary South

Dear Mrs. Kennedy,

I wanted to write long ago, but somehow I've never found the time. First, please let me introduce myself. My name is Elisabeth Zimmerman and I was born in Grenoble, France on April 3, 1951. I have never known what it is to be an American until November 22, 1963. Now, when the teacher calls on me in school and asks, "What is Nationalism?" I can describe it perfectly. November 22, 1963 is a date you don't have to write down to remember. I will remember it perfectly, forever.

It was around 1 o'clock in the afternoon and I was happily walking to the library when a negro boy approached me and asked, "Did you hear about the President? He was shot!" I merely nodded and thought he had some nerve thinking such bad thoughts about the President. I never so much as had an inkling, or maybe thought for a split second this maybe true. I had completely forgotten the President and Mrs. Kennedy had gone to Dallas but it wouldn't have made a difference because I didn't know what kind of a city it was. So I just continued on my way not giving a second thought to these very true words. But the second I opened the door to the library I knew he was right. It was true. The library's radio was on which it never is and was saying the President was having a transfusion, et cetera, et cetera. I just stood and thought of My President just lying down, surrounded with doctors and nurses trying to save his life but I couldn't. My knees felt weak and I quickly sank into a chair. I didn't cry, I couldn't. I could just run home to tell everybody. But everybody knew. Everybody was listening to the radio and I jointed them until around 1:30 p.m. when the news that changed everyone's life came.

THE PRESIDENT IS DEAD.

I went into my room and closed the door. I saw two, little children smiling up proudly at their parents. I saw my President making a speech and shaking his ever wagging finger. I saw my First Lady waving and smiling and I cried like I never cried before. I looked out the window at the unsuspecting people doing their usual things and I longed to call out "The President is dead. The President is dead. He's dead. He's dead!" But I only lifted my face toward heaven and I asked God one simple question.

WHY?

I don't know why, you don't know why, only G-D knows. But G-D doesn't want to tell. He made my President die an awful death. An unexpected, instant death. And the man who could have said so much was also killed. G-D made it very obvious that he didn't want his people to know.

WHY?

I have learned a great lesson. Don't take things for granted. I took it for granted John Fitzgerald Kennedy would be President again. I took it for granted that when I went to Washington this year I would maybe see my President and my First Lady and their children. I am in the 8th grade and in my school the 8th graders always go to Washington in around May. Now, I <u>might</u> see a President Johnson. But I don't want to, yet there's nothing I can do. I wish I had seen John Fitzgerald Kennedy so I could have something to remember him by. But I have only my magazines which I keep in a special drawer, "My President Kennedy Drawer" I call it. I know these magazines by heart, yet always I look at them and cry over them, and always I find something new. For the first time on November 26, 1963 I wished I had school. . . . I know I will <u>never</u> forget John Fitzgerald Kennedy. In school, I sit next to the window facing broadway. There is a big, white thing, a garage I think and on it is a star. Around a foot away in big red letters it says TEXACO, a name of a gasoline. You see now why I don't forget. I am dismissed from school at quarter to five and get home at around 5:30 p.m. One day, it was one of these lovely days. I was walking home from school towards River Side Drive where the Hudson

River is located. It happened to be snowing and it was beautiful. Suddenly, I stopped by a car. I looked around taking in the scene. Twilight. Sunset. the sun streaked with pink and violet. Tall street lamps intensifying the beautiful, white snowflakes against the darkness of this twilight. The snow was falling softly, frosting everything with an icing of white. The car I stopped by was covered with a thick blanket of this lustrous white. I acted on impulse. With my hand, I carved a little square and next to it a flicker of snow shaped like a flame. On the square I wrote J.F.K. and cleared the snow from around it. When you write about it, it is nothing. When you do it, it is something. I wrote mainly to say I love John F. Kennedy and when I stop to think, I find that I mourn for him more as a father of two lovely children and a husband of a charming woman more than as the President of the United States. . . .

Next time you visit The Grave please give the President my regards. Thank you. I can't wait to go to Washington being in the 8th grade. I often dream of being there and suddenly seeing you with Careline and John Jr. But it is a foolish dream. Yet still I dream. Every night before falling asleep I picture Careline and John Jr. sleeping peacefully. Then I picture my President lying, and a split second latler his grave with the eternal flame burning brightly. Then, I picture you Mrs. Kennedy, dressed in black with red, swollen eyes and I throw a kiss, and whisper "bon soir."

Yes, John Fitzgerald Kennedy is dead. But his memory is not and will live throughout history, forever.

Yours very affectionately,
Elisabeth Zimmerman

P.S. If you don't wish me to continue writing, please let me know.

As November 22 unfolded, Americans learned of the President's assassination in a myriad of settings. The circumstances in which they heard the news loomed very large in the minds of many from the start. In public places, walls between strangers tumbled down as reports of the President's death spread. On busy city sidewalks, in buses, taxicabs, department stores, and hospitals, citizens tried to absorb facts that seemed truly unfathomable. Patients in hospitals described receiving the news as they lay in their sick beds. "I don't know how to begin to tell you how I felt when one of the Nurse's Aide's came into our room and said have you heard?" wrote one hospitalized woman who was awaiting surgery. "The President has been shot! What startling words!!!!! Oh! God! no, I uttered. . . . We were sick in the hospital but we felt much worse with the President's passing." One man dying of cancer in a Veteran's hospital turned to his daughter and asked, "Why couldn't it have been me? He was so young." At college football practices, in small Alaskan Indian villages, in post offices, and along mail delivery routes, word of the President's death reached around the country, instantly halting the daily activities of millions.

BELLE HARBOR, L I.
N.Y.
NOV. 25TH 1963

Dear Mrs. Kennedy,

May God bless you today and always. My family joins me in a prayer for your well being at this sad time. I feel compelled to write to you and try to express my sorrow at the incalculable loss suffered by us all. Life's tragedies leave their ineffable marks on every human being, but the loss of our beloved President brought such deep, profound sorrow, we shall never forget. Words cannot express my emotions. I am frustrated and at a loss to convey to you the depths of my feelings.

The news came blasting at me from a woman with a transistor radio clutched to her ear, while shopping at Bloomingdales in New York City. Suddenly, strangers were strangers no longer. We turned to one another unbelieving and shocked. We shook our heads—this could not be! A crowd of us ran to the radio department and the world stopped still. The salespeople gathered with the customers, about the floor in trance like clusters—some sitting or down on their knees—all straining to catch every word. When what we dreaded came true, we were all crushed, defeated. I was reduced to an automaton—a sleep walker hoping to awaken soon from a horrible nightmare. Plans for shopping put aside, my walk down to the subway was unreal. The man who took my change for tokens; the news vendor; the conductor; the people young and old of all races, suddenly became united by the heavy burden we carried in our hearts. I saw groups of school children with bewildered grief written on their faces. Men and women wept openly and unashamedly. The same scenes prevailed on the bus I took to get to Belle Harbor. There was much silent meditation and prayers and those who spoke did so in hushed whispers—of your husband's goodness, his great leadership, his intellectual prowess his youth and his accomplishments.

At home, I found my children at the television set with tears in their eyes. We clung to one another for some one near and dear to us had passed on. We prayed for you and your family—for America—and for the world.

May God give you strength and many years. May your children and my children grow up unafraid and brave—in a better America because of your husband and the inspiring heritage he left us all. Amen

Mrs. Shirley Golub

DENVER, COLORADO
JANUARY 12, 1964

Mrs. John F. Kennedy:

Please allow me, a cab driver, to offer my condolences to you & your children.

The day this terrible thing happened—I think about 90% of the people of Denver cried a little bit—Everyone who got in my cab—People on the street—In the stores—patrons—clerks—cops—I saw people just standing on the corner—Just—Just unbelieving.

Well, I guess we can <u>all</u> shed tears—Mrs. Kennedy.

I might add—Since this tragic day I have heard so many comments on <u>how</u> <u>well Mrs.</u> <u>Kennedy</u> <u>stood up</u> <u>under</u> it <u>all.</u>

You have more strength than most of us American People! You are <u>admired</u> for this.

All I can say is—May God Bless you, "John John" & "Carolyn"

My heart goes out to you all—!

Why! Why! I'm still asking myself why—!

God Bless You & Yours!

Miller A. Alley

SATURDAY, NOV. 23, 1963

Dear Mrs. Kennedy;

Please forgive me for this intrusion during your mourning period of your very recent bereavement. It is with deep regret that I must write this letter, but in order to regain peace in mind, I feel that I am compelled to do so. As you read on, I hope that you will understand why.

First, I wish to introduce myself in a rather blunt way. My name, Leonard C. Rice, age 45,—occupation mail carrier,—military service, W.W. II, Korean & USN retired,—local community standing, plain ordinary citizen,—political ambition none,—other ambitions, to raise my family & to live & let live in peace.

Yesterday morning Friday Nov 22, as I was carrying my route, one of my patrons told me the tragic news that the President met with a tragedy. It came as a stunning shock to me, but as brutal as it may sound, not as a complete surprise.

After recovering from the feeling of nausea, my mind flashed back to a short time after Mr. Kennedy had become president-elect. My mother in law, mother & father & I were were discussing the prospects of our future president, who we helped elect in our humble way by casting our ballots for him. In our conversation, we agreed that Mr Kennedy would have a very difficult time with his new administration duties in these very trying times, and that he would be under fire during most of the time that he is in office. Also, in the conversation, I made the predictions that I now wish that I had kept my mouth shut & let well enough alone. I predicted that a strong attempt would be made to have Mr Kennedy impeached, or he would be a victim or attempted victim of an assassin. I then made the augury that if Mr Kennedy should survive these attempts through his term or terms in office, that he would go down in American history as one of the greatest presidents that our nation has ever had.

I now return to the time of yesterday, as the reports poured in from the people on my route (almost from house to house) on the Presidents condition, a prayer was on my lips as with millions of other Americans for his recovery. My prayer continued up to & beyond the time of the even more tragic news of his death, that he would survive recuperate & carry on his term of office, thus fulfilling the last part of prophecy.

After returning home & listening to the news reports & then spending a restless night still clinging to the hope it was all a nightmare, I was confronted with the blazing headlines of the morning paper, only then was I fully aware of the awful tragedy.

I know now that my prayers were not in vain. for Mr. Kennedy was a very great man and as truly a great president. I am proud, just as millions of other Americans must be, to have served under a president with courage beyond compare.

> Respectfully yours,
> Leonard C. Rice

Dear Mrs. Kennedy,

I feel a great need to express to you the shock, dismay and grief that I, my family and everyone I know, felt at your husband's tragic death. Even now, months later, it is impossible to view his name or his picture without a lump coming into my throat. An old Peter Lawford motion picture on television will remind me of your husband, and I cannot bear to watch.

My husband and I had looked forward to November 22 for almost a year. We had gone through all the red tape of adopting a wonderful little boy from the Los Angeles County Bureau of Adoptions, and the final steps to make him our legal son were to be taken on November 22. All our thoughts and efforts that week were directed to our morning in court which was to be a cause for celebration.

We were not even aware of the President's trip to Texas. Sitting in court in our best clothes, waiting to be called into the judge's chambers we heard the first report of the shooting from a late arriving lawyer. We were shocked and though we took our turn in the judge's chambers and our son became our heir, our thoughts were on the President and you.

In our lawyer's office as we directed him regarding the drawing up of our wills, we heard the unconfirmed news of the President's death. We hurriedly finished our business with the lawyer and dashed home where we learned the final truth from the television.

My husband is normally an undemonstrative man. He was on the verge of tears.

We lived in a vacuum that week end. Watching television, unable to escape one minute from the tragedy, it was worse than a family death for us. And many have said publicly, your dignity and self control were majestic. We were very proud of you and of your children, the sight of whom broke our hearts.

When your official year of mourning is over, I hope that you will return to some facet of public life. You are qualified to enter so many fields—art, journalism, government, fashion, etc. You truly have become a symbol and a goal for American womanhood with your sense of beauty, dignity and grace.

This letter is most inadequate, but it is my only way to express my deep concerns for you and your family.

Sincerely,

Janeen Ostby

UNIVERSITY OF MASSACHUSETTS

Dear Mrs. Kennedy:

I have never seen our football players cry . . . but today, they did.

Martin Rosenberg '65

U of Mass

CUSTER, MICHIGAN
NOVEMBER 23, 1963

Most Gracious Lady, Jackie and children,

May I, one of the millions of little people dedicated to serving under our President, and a clerk in a small third class office of the Postal Department extend my deepest sympathy to you and yours at this time.

At the announcement of the tragedy my associate and I immediately began a silent prayer that our beloved leader would live. When the announcement of his death came it was my duty to lower our flag to half mast.

I can not explain the emotion I felt and the saddness in my heart at our loss as I lowered the Flag. Never even with the loss of several of my family have I been more deeply touched than with the loss of President Kennedy.

I only pray that this life taken so unjustly has not gone in vain and our nation survive the loss.

Sincerely,

Mrs. Frances Nash

NOVEMBER 22, 1963
ST. LUKE'S MISSION
SHAGELUK, ALASKA

Dear Mrs. Kennedy:

I want to convey to you my own personal grief at the news of the death of the President. The news came as an unbelievable shock and brought me to my knees with a prayer for your husband and you and your children.

This evening at seven our parish hall was full as we prayed and celebrated a Requiem Eucharist for the President. During the day I visited most of the homes in our small Indian village and everywhere I went I could see the deep sorrow and concern everyone feels at this time. Several people walked over two miles at ten degrees below zero to attend the Requiem. Your husband is loved and respected by the people of Shageluk.

Please know that you and your children shall be constantly in our prayers in the days to come.

God bless you.

> Faithfully in our Lord,
> The Rev. David Keller
> Priest-in-charge

NOVEMBER 23, 1963

Dear Mrs J. F. Kennedy & family

We send our most sincere sympathy. It seems as though a part of us all has died and I believe most others feel the same.

We have four girls (Lisa Marie 4½ , Monique 3½, Michelle 2½, Je'Neanne 1 and one expected January 19th). The evening of Friday November 22nd and in a May Co. window our oldest stopped fast— there were a group of people standing in front of a picture of President J. F. Kennedy with his birth and death dates. She pressed her face up against the window and looked at the picture for what seemed like a very long time. She turned and looked at us. (I hadn't realized at the time that she had been paying attention to the T.V. announcements during the day but apparently she had). I believe our little one expressed the thoughts of millions although only 4½ she came out with a child's sincere thoughts that we will never forget and am sure all those present won't either.

"Mama, look at his face—he's so good—maybe God didn't think we loved him enough. Why did someone shoot and hurt him?"

By the time she had finished, tear stained faces turned our way—she looked for an answer—I was so taken with her thoughts that I just shook my head and said "I don't know" and tried holding back the tears. One lady stepped up in front of us and said "How old is she?" I told her 4½. "If a child of 4½ can feel that way—then how much more should we grown ups <u>really think</u> and feel responsible for the horrible thing that has happened in our country."

He tried to do so much for so many—"maybe God didn't think we loved him enough"

May the sympathy of your friends help you throughout your sorrow. May God give you hope and courage to meet each new tomorrow.

With our most sincere sympathy,

Mr & Mrs. Roland A. Fiola

Lisa, Monique, Michelle and Je'Neanne

Fiola

(My husband was born in Fall River, Mass) and we know how the people of that area must feel. May God bless you and yours through both tradgies suffered this year.

Dear Mrs. Kennedy—

Enclosed is a copy of a letter written by our son, Larry Jackson.

I believe that it is self-explanatory; an expression of grief from one young American.

May I offer our personal salute to you, Mrs. Kennedy, for your singular courage and quiet fortitude during these dark days? You are a splendid example to us all.

Sincerely,

Mrs. Whitley Ray

Saint Mary's College
California
November 22, 1963

Dear Mom and Dad,

Why?

He towered above them all

He exuded greatness, overshadowing opposition

He had vitality, drive, ambition, charm

He had wisdom, control, maturity, decisiveness

He had love, for God and men

He took over the helm of the nation

He challenged the people for their responsibility

He led them all in accepting it

He stood for justice, truth, and liberty

He resisted ignorance, hate, and apathy

He astounded with patience and courage

He spoke and was heard

He commanded and was obeyed

He loved and was loved

He did what he knew was right

He was hailed by a hopeful world shouting cries of "Kennedy! Kennedy!"

His life was precious to all

And so they shot him.

Again, why? The whole thing is too horrible and shocking to believe. I'm not angry—I'm sick. I don't want blood—I want an answer. I don't want to kill—I just want to cry.

This is my hope. That Kennedy, like Christ, is love struck down by hate; and that, in a way like Christ, he will rise again from the coldness of death to which the forces of hate seem to have damned him. His spirit will live on with his followers to defeat these enemies of humanity, tearing them down from their pedestal of petty triumph and hate and injustice, but I hope that this great sacrifice succeeds in somehow lessening their effects on us. We cannot make him into a martyr because that's just what he is. Bewilderedly, we ask the question "Why?" Let's hope that there is an answer. Let's hope he did not die without cause. Such a thought is unbearable.

College students are not without feelings. Their interests were pretty apparent today. One hour after the news of his death, the chapel was packed with an unprecedented amount of students for noon Mass. Stunned silence reigned over the campus and people walked around with glazed looks. Red eyes were not infrequent. The flag flew at half-mast while the SMC on the hillside was reverently changed to the letters JFK. The announcement at an unusually quiet lunch that classes were cancelled for the rest of the day drew no cheers. This was not empty sentimentality, it was really deep feeling.

I am looking forward to coming home and seeing you all, hoping that these expectations will clear away the depressed mood I'm in. It's raining now; I know that it's silly and it's been raining on and off for the past few weeks, but I can't help feeling the world is crying.

Love,
Larry

Some who received news of JFK's assassination were old enough to remember the assassination of other Presidents. Three previous American Presidents had died in office from an assassin's bullet—Abraham Lincoln (1865), James Garfield (1881), and William McKinley (1901)—tragedies that fell within the life experience of several elderly Americans in 1963. As one ninety-year-old Californian wrote to John F. Kennedy Jr., "I was born the first day of April 1873 and your father is the third president of the U.S.A. that has been killed in my time. I was around 6 years old when Garfield was killed and about 21 when McKinley was killed and I was 90 years and 8 months old when your father was torn from us."

Kennedy's death likewise evoked memories of the last President to die in office—Franklin Roosevelt. Letter writers recalled how shocked they were on April 12, 1945, when word came of FDR's death. The most common point of historical reference in the condolence letters, however, was the assassination—or the "martyrdom," as many put it—of Lincoln. No previous assassination of a President took place, however, in a period when mass communication permitted such rapid and wide access to news, images, and analysis of the event.

Many stressed that Kennedy's youth and vitality, as well as the assassination's ghastly circumstances, made his death especially harrowing. "I am old enough," one retired navy captain wrote to Mrs Kennedy, "to have heard the moan of the nation upon the death of President McKinley and I remember well the angry, mournful growl which rose from its throat following disastrous Pearl Harbor, but never have I seen nor heard such overwhelming national grief, nor do I believe has anyone." "We have just seen," he observed, "the strongest nation on earth brought sobbing to its knees in abject grief for the first time in living man's memory."

CAMDEN, W. VA.
DEC. 1, 1963

Dear Mrs. Kennedy

I want to express my sympathy in your great loss, and in this trying hour of our Nation. It is with sorrow I have to say I have seen all four of our Presidents assassinated, as I Celebrated my 99th birthday Nov. 22. It Certainly was a very sad evening for me, as well all West Virginians.

We had learnd to love the President, as he was so interested in our state. The Nation has lost a great leader.

May God bless you and the Children is my prayer.

Sincerely
Perry C. Gum

....❋

CLEVELAND HEIGHTS, OHIO
NOVEMBER 1963

Dear Mrs. Kennedy,

Your bitter experiences of the past few days should never have happened, but it did. It has happened to an infinite number of persons who have unexpectedly been made widows and fatherless. It has happened to very few families who had previously given their husbands and fathers to the lonely responsibility of the Presidency.

When I was a child my Grandfather told me of the dreadful loss of the man who was both his father and President of the United States. At the time of his father's assassination in 1881 my Grandfather was fifteen years of age. Perhaps your children are blessed by not yet realizing their terrible double loss. Losing a parent so prematurely as many of us have is a great loss, but the loss of one who is loved and respected by so many for such remarkable capacities apart from the personal family relationships is perhaps even more acute. Your children will come to know him through you.

Only now have I come to have some realization of the great personal loss which Grandmother Garfield felt at the time of her husband's death. No longer is it only an isolated fact in history. Assassination is merely a dry euphemism which applies to the outright murder of so prominent a person. The terminology does not lessen the pain of those who are involved at the time. This has been a personal loss to us all.

I write to extend our sincere wishes for your continued strength beyond the immediate trial to which you have been subjected. The tragic happenings of the past few hours have made more obvious to each of us the supreme gift a man must make in his acceptance of such responsibilities. A consequence so violent and unnecessary seems impossible in our present day existence, but I fear we have made less progress than we believed. God willing, the price exacted from you and your family will in some way contribute to a greater maturity and selflessness in all of us and to the realization of the peace and understanding your husband sought for his country.

Your remarkable dignity and strength during this time have been a great lesson to us all. You have our deepest affection and greatest admiration.

Rudolph H. Garfield

Dear Mrs. Kennedy:

Tonight, in the homes of the millions of average families who make up this nation, virtually every thought is of you, your children and your sorrow. Many of us wish we could find some words to console you and let you know that, to whatever extent it is possible, we share you grief. You lost your husband; we lost our president and our leader.

Perhaps, since your husband was a student of history, the phrases which come closest to saying what is on the heart of this family tonight were written by another president, also martyred, to another wife and mother: Abraham Lincoln in his letter to Mrs. Bixby. In it, he wrote in part:

> "I feel how weak and fruitless must be any words of mine which should attempt to beguile you from the grief of a loss so overwhelming. . . . .
>
> "I pray that our Heavenly Father may assuage the anguish of your bereavement, and leave you only the cherished memory of the loved and lost, and the solemn pride

that must be yours to have laid so costly a sacrifice upon the altar of freedom."

Our prayers, our thoughts and our wishes are with you and yours.

Sincerely,
Fred R. Zepp

NOVEMBER 23, 1963
DOVER, DELAWARE

Dear Mrs. Kennedy,

What can anyone say at a time like this, only that I'm sorry. I was ten years old when President McKinley was killed and I never thought I would live through a second Presidential Assassination, in this wonderful country of ours.

God Bless you and your family and take care of yourself.

Very truly yours,
Mrs. Regina Metzger

ILLINOIS
NOVEMBER 27, 1963

Dear Mrs. Kennedy and family,

Finding the right words of comfort at a time like this is always difficult because all the words in the world won't bring a loved one back. His laugh will never again be heard, his smile, seen or his hand, felt. In another way, however, he will never die. In our religion we say that as long as a person is remembered, he is still living; therefore, John F. Kennedy will never die. He will live on in the thoughts of those who knew and loved him. He will live on in history, never to be forgotten.

No one will forget what he was doing on Friday, November 22, 1963, when he heard the news that John F. Kenned had been shot. No one will forget what seemed like the hours we waited until we heard the final, awful news that he was dead. While watching the television show "As the World Turns" the news announcement came. While waiting for the announcer to say what the news was, I thought to myself, "What election results in some outlying location is he going to tell us this time." When he finally said that the President had been shot I stared in disbelief at the set. There must be some mistake. This doesn't happen in the United States. It doesn't happen now. This happens only in history books.

My thoughts immediately went back to another day in 1945, when I was nine years old, and listening to "Terry and the Pirates"; and the announcer broke in to say that Franklin Roosevelt had died. I remember going down to tell Mother, who was in the kitchen making dinner. She told me that it must have been part of the story. Finally I convinced her to turn on the radio and she heard the news for herself and began to cry. At the time I wondered why. Now I know. For I cried Friday.

With the advent of television the average person is even better acquainted with the President than he was in 1945. The President is seen regularly and one can't help but feel that he is a member of one's own family or that of a next door neighbor. John F. Kennedy was more than a politician giving speeches. He was a human being with a wife and family. The people knew him as a wonderful father, husband, brother, and son. After the initial news that the President had been shot, I was glued to the television set until the phone rang. While talking I happened to look out and noticed that the flag at the school, which can be seen from the kitchen window, had been lowered. At the same time it was announced on television that John F. Kennedy was dead. My heart sunk.

The rest of the afternoon I, like many others, was in a daze. The baby had to be dressed to go to visit her grandmother and I, too, had to dress. I really didn't feel like going, but there was no way out of it. It was raining outside and the traffic was heavy. However, even though it was the rush hour, no one seemed to be rushing. There wasn't the usual screeching of brakes and honking of horns. Everyone was listening to the radio and in a state of shock. The drive took twice as long as it usually does, but no one seemed to be conscious of it. Dinner was eaten and my husband, the baby and I went home. The radio was on during the whole ride. Usually it is not. But the day hadn't been a normal one. John F. Kennedy had been killed. When we got home, the television was turned on; and we watched transfixed before it. It remained on most of our waking hours until late Monday night....

Each person wished that there was something that he or she could do for the grieving family. Why worry about the nation? Luckily we live in America where when one person dies our whole political system is not thrown up for grabs. There are no new forces to take over the government. It will go on pretty much as it has always gone on. But the family of John F. Kennedy will never be the same. You've lost a husband, a father, a brother, a son, a cousin, or a very close relative. And we, the people, have lost a friend. Our sadness lies not in the fact that we have lost a friend, but, because we feel that, in a way, we are all responsible for your loss; and there is nothing we can do about it. There is nothing we can say. But we must do something. So we enclose this money to be used for the benefit of retarded children or in any other way that you think best. It is the least that we can do.

Our hearts cry for you and for your family. You have shown us and the world the dignity and the strength you have. We hope you know our thoughts are with you. You are not alone in your grief. If when we turn on the radio and we are offended by the noise and the uselessness of the regular programs, commercials, and music; we can imagine how you must feel. We can't seem to concentrate on anything, it all seems too trivial. But we must continue because life must go on for the living. That is what John F. Kennedy would have wanted. He wouldn't want us all to collapse because of his death. As a tribute to him we must try to make the world a better place to live so that this type of thing will never have to happen again. We hope that in some way this letter shows you how so many of us feel and will, in some small way, help you in picking up the threads of your lives and weaving them back together.

Sincerely,

The Howard A. Jacobs family

Many Americans who were overseas felt an acute sense of dislocation when they learned of President Kennedy's assassination. Those who wrote to loved ones in the United States described not only the specific circumstances in which they heard the news but also the reaction in the far-flung places where they found themselves on November 22 and in the immediate days following. The raw emotion and painstaking detail in these letters deeply impressed their families, many of whom enclosed these messages in their own condolence notes to Jacqueline Kennedy.

JAN. 25, 1964

Dear Mrs. Kennedy,

My husband sent me this letter when he was in Trondheim Norway, on Nov. 22nd when he was on tour with the Los Angeles Chamber Orchestra.

He expressed in words what I felt in my heart & I feel I would like to share this with you.

With most heartfelt wishes for you & your family,

Respectfully,
Katherine Stamos
Torrance, California

Trondheim, Den Nov. 22 1963

Time 10:30 P.M.

My dearest wife—

This is one of the darkest moments of my life—tonight as we were awaiting Henry to come on-stage between the pieces during a concert—there was a great delay—then finally he and Jackie came on and started an aria—we played about 1 minute then he stopped and said "I can't go on any more— President Kennedy has been assassinated."—then next thing I knew I was crying, as if I lost my own brother or father—but more than our own personal lost—we have lost our leader, our hope for a better world within our own lifetime. They did the same thing to Christ, to Lincoln—and now Kennedy—may God help us in this hour of great darkness—our faith must be strong as Kennedy's was—he took the chance—to try and reform the hidden ignorance, the bigotry of the past and now he has become a martyr. He has touched our own personal lives. <u>I would not be here if it weren't for Kennedy</u>, trying to show the world that we are a cultured people. But some of us aren't—as long as we have fanatacism to have produced such a dastardly deed.

You remember when we first came to Dallas, the day after he was elected—the place was in mourning—I always sensed that ignorance was somewhere nearby—it is quite ironic that this should be scene—the modern day "Calvary"—at least he died near his wife—and as Senator Mansfield said "He has gone to his reward"—may his example serve as an inspiration to each and every one of us—may we never forget what he fought for—equal-rights for <u>all</u> men, peaceful co-existence and like Roosevelt's dream of a one world nation

I feel that these things must be said, and only to you my dearest wife because it is in moments such as these that only you understand—I never felt so lonely, so much in need of you, to have you nearby—than now—we are all stricken—we all wished we were home—the news as it comes over the radio has to be translated to us.

I pray, and we must all pray, that his nation shall triumph no matter how many crosses we have to bear—and this is truly a heavy-one indeed—because without faith we are lost—this country was founded on faith. And we must each try and carry on these ideals in our own lives—that's the way men like Jack Kennedy hoped for the people of his country.

Pardon this letter of thinking out loud—I know how you must feel—keep up your spirit—I'll be home soon—God bless you and everyone there

Your husband. Spiro

••• ❈

CHRISTMAS 1963

Dear Mrs. Kennedy

I do not know if you will actually see this yourself but I hope very much that you will read the enclosed copy of a letter I received from my daughter who is teaching English in the Peace Corps in Ethiopia. She expresses what we all feel and I think you would like to read her letter.

I wish you and your wonderful children the very best. I have the greatest admiration for you and know that you will make a fine new life for yourself and your family.

Very sincerely,

(Mrs. John) Anna Lounsbery

November 23

Dear Mom—

Even the Ethiopian sky is in mourning today, and it was raining last night when President Kennedy died. We were having a housewarming party for the nurses when Ron came in with the news. We didn't believe him until he turned on the radio—we sat in silent horror and each time the words were repeated it was a new shock—and still is today, and will be for a long time to come, for presidents aren't assassinated in the modern world. I remember that you cried when Roosevelt died, but of course it meant nothing to me. I feel now as if a member of my family had died. In a very real sense he was our idol; he is the reason for our being here—his idealism, his courage. It is difficult being away from home at a time like this, yet in a way, it is even more meaningful, for we can sense and see the deep sorrow and respect for this great man felt by foreigners—people who never knew him, but know of and about him. Flags are flying at half mast, radios are playing masses and requiems, our Ethiopian friends have come quietly to express their grief, His and Her Highnesses had an audience with us to show their sorrow. The irony of Lincoln, first champion of civil rights and equality, being killed by a maniac, and now Kennedy—so much like him, both being succeeded by Johnson, a hundred years ago it was called the Reconstruction period, and indeed, that is what the years to come will be—a period to reconstruct and attempt to follow the patterns of thought and action so vigorously set down by Kennedy. Truly, the greatest tribute anyone could pay to this man is to shake off complacency, examine one's own heart and mind, and dedicate oneself to the cause of peace, and the eradication of poverty, disease, and inequality. The world is hurt, angry, and weeps. Shall he have lived and died in vain; shall he be a martyr? Surely history will be changed—I pray that we may all learn from him, and work better because of him.

With Love, Annie

16, JAN. 1964
PITTSBURGH, PA 15227

My dear Mrs. Kennedy;

I realize with the tremendous amount of mail you are receiving—it is impossible for you to read it all.

Never the less—I sincerely hope you will be able to read this one.

Last Nov. 30, we received a letter from our son, who is in the U.S. Army Signal Corps—stationed in Mannheim Germany.

We think it is a beautiful letter, and would like to share it with you.

When we first received it—I <u>considered</u> sending it to you—but didn't.

Now, at the suggestion of <u>many</u> friends & neighbors who have read it—I am sending you a copy. (I just <u>cant</u> part with the original letter!).

May God Bless You And Your Family,

Respectully,

Mrs. M. F. Melder

Mannheim, Germany

26, Nov. 1963

Dear Mom and Dad,

Now that all the tragedy is over—or at least I pray to the Almighty Spirit that it is over—I begin a letter of condolence. And that is as it should be; for we are Americans.

Any eulogy that I may attempt will fall short of those already made, and also far short of what is necessary. He is dead. The President is dead, and he has just been buried. He now becomes history. I am crying now because my President is dead, final and forever. If he had not gone to Dallas; if he'd been sick that day; if it had rained the bubble would have been up. If, if . . . we might have had his leadership a little longer. If a cruel, cool, selfish animal had not wanted to rob us of our hope and our temperament, then we would not know this loss.

I had the good fortune to see the funeral walk to the church via Telstar in Christa's home and I felt pride. Proud of all the world's leaders paying tribute; proud of the bright sunny day in our Nation's Capitol; proud of the more than three-quarters of a million Americans paying silent tribute, and proud; deeply proud of his widow, herself proud and strident in grief; as it should be, for she is of a noble family.

We did not see the funeral service, or anything that followed. We were then in Berlin, in a massive square; acres of faces, some young and alert, some old and crying, some straining to see, some with their heads bowed and all quiet, all these many, many people. Two speakers spoke and although I do not speak good German, I understood. I understood every pause, every choke, and every gesture. I understood the signs in the crowd, reading "Wir haben einen gute freund verloren" (We have lost a good friend), and "J.F.K. er war eine Berliner" (J.F.K. was a Berliner), and I understood the speakers restating the phrase, "Ich bin eine Berliner"—with feeling and grief. A silent large photograph of him, taken during his famous trip, being over the scene and then the speakers stopped, and the camera switched rapidly to an American soldier, in full dress. He slowly lifted the bugle to his lips and played taps, and I was proud. My chin was high and the salt tears flowed, as they are now. Then all the lights went out, and then came suddenly on, lighting up this huge square as John F. Kennedy Platz—which I Will See. An hour later they lit the eternal light over his grave, lighting a way to freedom, and to peace.

This whole hideous thing need not have happened. But it did and we, too, are a part of this history; part of its horrible beginning and part of its peaceful end.

Love, Ray

Lt. Raymond A. Melder

As Americans absorbed the shock of President Kennedy's assassination, the President's body was brought back to Washington, D.C., on Air Force One, leaving Dallas less than two hours and twenty minutes after the assassination. Vice President Lyndon Johnson took the oath of office as the nation's thirty-sixth President aboard the aircraft before its departure. The return of Air Force One to Andrews Air Force Base near 6 p.m. (EST) was televised live, and it was a shocking scene for many observers. Jacqueline Kennedy deplaned wearing the same pink suit she had worn in Dallas, her skirt, jacket, and stockings visibly smeared with her husband's blood. Throughout the day she had rejected entreaties from those who urged her to change her blood-stained clothing, especially as she stood next to President Johnson during his swearing in. Ignoring a white dress laid out for her on Air Force One, which she expected to wear on the last part of their trip to Austin, she asked herself why she would "want the blood off?" "Let them see what they've done," she later remembered concluding. Now accompanied by Robert Kennedy, who awaited Air Force One's arrival in Washington, she rode in an ambulance with her husband's body to Bethesda Naval Hospital, where an autopsy was performed.

In truth, a sense of guilt and responsibility settled quickly upon more than a few Americans within hours of the President's assassination. Acutely conscious that much attention was focused on Dallas, those who lived in the city expressed particular anguish in letters written on the very day of the assassination. Several anticipated that Texans would be blamed for Kennedy's murder and expressed sadness at the judgment they expected to be leveled. "Most persons outside the state will bitterly feel that it could only have happened in Texas," one woman from Midland predicted. Others internalized this expectation and felt terrible remorse. A few Americans imagined that their own actions, or inaction, on the day or night before the President's death—no matter how small and seemingly unrelated to the tragedy—had caused the murder of the President. "I feel in some way I could have been some what at fault," one individual explained. "Each night I use to pray for your husband to protect us with God's help. Last night I stayed late went to bed without saying my prayer. When I heard this about our great leader I was so sorry I did not pray. This I will never do again. God help me."

NOVEMBER 22, 1963
DALLAS, TEXAS

My dear Mrs. Kennedy:

I have never before written to a Congressman, President or any type of Statesman. In fact, in my thirty some years of living I have never DONE MUCH OF ANYTHING, except vote, toward being an American or making this Country a better place in which to live.

Today, however, my heart is SO HEAVY I feel I must express myself to you. I feel I must tell you how VERY ASHAMED I am to be living in this city. Dallas—a city of cultural background—a city of colleges, schools and supposedly intelligent people. God would that I could move from this place this hour.

I happened to be downtown this noon, and took the time from my work to stand on the street to take a look at a man and woman that I have LOVED for 3 years. I am SO GLAD now that I did, for never in my life have I admired and respected a man more than I did (and do) JOHN F. KENNEDY. In my opinion, he was the most outstanding individual that this Country has been able to produce in hundreds of years.

All of this means nothing to you now, I am sure of that, but my dear lady—I was moved to tell you that there are of us in this terrible city of ours who LOVED your husband very much. We have cried (literally) bitter tears over this day. May God forgive those who brought this shame to our city—to our Country.

I would like to extend to you, Mr. Kennedy's family, your family the most sincere and heart-felt sorrow that is felt by me, my Mother, and many thousands of others in this hour. I only wish it were possible for a humble, small person such as I to bring this message of sympathy to you in person. Of course this is not possible, but I pray that my letter DOES reach you personally.

> Most Sincerely Yours,
> Robert L. Wood

DENTON, TEXAS
1:10 P.M.
NOV. 22, 1963

From a student of North Texas State University
The radio sat in the window of the second floor dorm window blaring out the sad news that our President had been shot! People

walking around in twos and threes stopped their happy chattering and stood silently on the street, waiting—listening—wondering—praying. It was a beautiful fall day—it was almost like spring—so balmy and lovely was the day, but a glum palor hung over the campus of NTSU. Cars drove slowly by, some stopped to listen, classes were excused, all activity on NTSU campus halted as we waited. We had thoughts only of our President. All Texans are ashamed of Dallas. God help our President.

He is dead! God receive his soul.

Mrs. Kennedy, I love you and I will pray for you.

God Bless You,
Eileen Mitchell

- ✳

NOVEMBER 22, 1963
LOWRY AIR FORCE BASE
COLORDA

Dear Mrs. Kennedy

I know this is a time of great stress to you and your family.

I do ask of you to for give my poor attempt of expression.

I am an airman receiving training at Lowry Air Force Base, Colorda. I have served under your husband for four mounth before this tragic happening. The men in my Barracks feel as I do of you and your late husband.

At the news of his death there was a emence silance through out the barracks everyone was sick in side. This sickness was noticable in everyone upon this base. The sickness ranged amoung airmen and officers alike. All showing the stress of the happening.

Mrs Kennedy, the reason I feel that I sould write is because the happenings of today is thirty miles from my home. I feel great stress up on myself it being so close to my home. It has made a great impression that will last my life time.

I feel If I could be a portion of the man your husband was I would thank God

If I could be as fortunate as your husband to have a wife as you, to be able to stand by me and give me strength and willingness that is needed by a man to be able to make the decision of life, love and happiness of millions of Americans. I would be the greatest man that has come along in many years.

You mean as much, as he did. Mrs Kennedy now you stand strong an pure in the eyes of American and God. upon you you have a strength that many men would give everything they have to possest it. I admire your great strength in time of stress and pray to God it will be as easy as possible upon you and your Family. I pray to God your strength will stand as a simbole to American women.

God be with us that we can make of this country what your husband wished it to be. Our prayers are with you and your family.

Mrs Kennedy as myself speaking for everyone in my Barracks and upon Lowry Air Force Base. Through the President John F. Kennedy this country is untied in God to form a strong and better union that will resist communism and all opression that is brought agest it. Through the love of God and my country I write my most Greaving letter of the most Great man who ever lived.

> Airman third class
> Kenneth R. Wiggs Jr.
> Lowry Air Force Base

YOUTH FOR KENNEDY-JOHNSON
AUSTIN, TEXAS 78723
NOVEMBER 25, 1963

Dear Mrs. Kennedy,

There are no words in any language to express truly our grief and the sympathy we wish to extend to you and your family on the death of your husband, the President—our President. We Texans pride ourselves in our state, and that such a perverted act would happen here doubles the weight of grief in our hearts. We in the organization have our own personal sorrows, too, for we hold a seemingly stronger bond to Mr. Kennedy than others.

Here in Texas—here in our city the sunsets are blood red and the streets flow with tears. The only sound heard is that of crystal silence, rippled occasionally by a single bell that tolls a solemn requiem for our President John F. Kennedy. Our prayers are with you, your small children, and the other members of your family, and also the deepest hope that somewhere in your private heart you can find that enabling you to forgive the twisted mind, to forgive Texas, and to forgive us, the nation.

In deepest sympathy,
The Youth for Kennedy-Johnson
Marcy Wentworth
Deputy Corresponding Secy.

EL PASO, TEXAS
DEC. 8, 1963

Dear Mrs. Kennedy:

I am but a humble postman and I realize the many letters you have received, which is but deserving to you, throughout this wide world. We at our house have continued to mourn the great loss to all of us, we are my wife and I and our four children, our youngest boy also three as yours. President Kennedy was adored in our house by all of us here and the small glimpse we saw of him in June when he visited here, shall always be remembered. We were still new here in El Paso and we saw the great love people here hath for him. We were both of the same age, I too having been born in 1917, we both were in the South Pacific area during the last war, but I was no hero as for him. We both married in 1953 and my wife is the same age as you are, we were married in Mexico City. We are also Catholics and had four children, two boys and two girls of which God was so good to us. I am not ashamed to say how terrible we all felt at this tragedy and even my five year old girl cried that day.

So our heart goes out to you and your very dear children for always. We will continue to admire your courage in the face of it all and continue to admire the great love President Kennedy had for the world. Please try to find it in your heart that we Texans of Mexican descent had a great love for all of you. We do hope that you will not think all of us Texans were bad, there is bad in every sort of people as you well know.

We will continue to pray for you and your family in the days and years to come.

May God bless all of you.

Yours most sincerely,

Henry Gonzales

P.S. Being a stamp collector, do sincerely hope the new stamp in memory of your husband will do him justice.

DALLAS, TEXAS
FEB. 7, 1964

MRS. JOHN F. KENNEDY
WASHINGTON, D.C.

Dear Mrs. Kennedy:

I have tried many times to put into words, my deep sorrow at the tragic passing of President John F. Kennedy. There are no adequate words, but I want to express my deep regret to you.

I feel sure, you have received many letters from Dallas. I know of the shock, horror, and following grief that took place here. We have not, and will never forget that he was cowardly and visciously taken from us here in our town. I know it is hard to understand in the rest of the nation, but I believe our grief is greater. It would have to be endured to be understood. God help us that it might never again happen to one so defenseless, but without fear.

Not all Dallas agrees politically. But many, many Dallasites were loyal and devoted supporters of President Kennedy. Those who did not agree always, respected and admired his great intellect and courage. He was our President.

I, cannot now, visit his resting place, but I pass the place he left us often, and each time, I say a prayer of thanks for what he did for our Nation, and a prayer that you and your children may walk in the shelter of God's love, and that time will stop the hurt, and bring you peace and happiness.

Most sincerely,
Mrs. J. M. Thornhill

BROOKLYN, N. Y.
JAN. 14, 1964

Dear Mrs. Kennedy:

I have wanted somehow to express my heartfelt sympathy to you.

I realize that sheer words may not be sufficient, and yet, this is the very way your husband has endeared himself to all his people. Words, which while he lived expressed his deepest convictions. Words that now will always be remembered in all who heard them, and for the future generations, in history's pages.

Your grief was shared wholly by myself and my family.

I am a patriotic person, but not until President Kennedy's death, did I own a large flag to fly.

I burnt a candle in his memory when you lit the "Eternal Light."

I admire you as a woman, mother and first lady of our land. I, fervently hope the future may possibly bring so fine and great a lady to your position again. I doubt though that two such gracious people could possibly live in my time.

My greatest dedication to your husband, is my personal keeping of all the events occurring from that infamous day on. Even though I have a son, I have begun a scrap book with a dedication to my young 6 year old daughter, Lorraine, for a peculiar event occurred on Nov. 22nd.

The dedication reads as follows:

> "I dedicate this collection to my 6 year old daughter, Lorraine, for the ironic and coincidental question she posed to me on this day November 22nd, 1963.
>
> On the way to school at 12:30 P.M., she looked up at the school flag and asked, "Mommy, when do we keep the flag at half mast?" I, in turn, explained when. She then said, "That means even when President Kennedy dies? . . .

I need not explain further, the eerie feeling that befell me later that day, when the news was out. Also the child's reaction was terrible. She, poor thing, felt responsible because of her question.

Closing this letter will not close my memory or any good citizens'. Close it I must, with the hope that "God," may smooth the future way for you and your children and keep you all from further personal grief.

My deepest Sentiments,

> Sincerely,
>
> Mrs. G. Katzberg

.... ❧

With Saturday's first dawn, newspapers published in bold headlines accounts of an event that virtually no sentient person in the United States was unaware of. "**KENNEDY IS KILLED BY SNIPER AS HE RIDES IN CAR IN DALLAS; JOHNSON SWORN IN ON PLANE,**" *reported the* New York Times. "**KENNEDY SLAIN ON DALLAS STREET**," *read the banner in the* Dallas Morning News *the day after the assassination. Mrs. Kennedy had returned at 4:30 Saturday morning to the White House with President Kennedy's coffin. It now lay on a catafalque, a replica of the one used after Lincoln's assassination, in the East Room. Preparations for a historic state funeral had already begun. Contrary to subsequently published accounts, Mrs. Kennedy did not singlehandedly direct these arrange-*

ments but made decisions that needed to be made along with Robert Kennedy, close advisers to the President, White House staff, and military personnel.

On Sunday, President Kennedy's body was to be moved to the Capitol Rotunda by a horse-drawn caisson. Americans who watched their televisions that morning, awaiting the solemn journey, first witnessed Dallas nightclub owner Jack Ruby murder Lee Harvey Oswald on live television. The event seemed utterly surreal and almost too much to absorb for those still reeling from Kennedy's assassination less than forty-eight hours earlier.

Meanwhile, many other Americans had embarked on a pilgrimage to the nation's capital. Huge crowds stood alongside Pennsylvania Avenue watching in near silence on Sunday afternoon as the caisson bearing President Kennedy's body moved down Pennsylvania Avenue from the White House toward the Capitol. Followed by a riderless horse, and a procession of cars bearing the Kennedy family, President Johnson, and other officials, the caisson commanded the attention of onlookers who stared at the flag-draped coffin. Only the sound of muffled drums pierced the solemn quiet of the moment. For the rest of the day and into the early hours of the following morning, thousands stood in line, some for ten hours, waiting to file through the Capitol Rotunda to pay their respects to the slain President. All through the night, the slow procession continued until the doors were closed at 9:00 a.m. in preparation for the state funeral. Monday was a national day of mourning. Most Americans observed it by watching on television the funeral procession from the Capitol to St. Matthew's Cathedral. Walking behind the horse-drawn caisson that carried President Kennedy's body now from the Capitol to St. Matthew's Cathedral, Jacqueline Kennedy and the President's two brothers led a delegation of mourners that included heads of states and dignitaries from around the globe. Their exposure as they marched seemed an act of defiance, given the reason for their gathering.

The terrible circumstances of John Kennedy's death, the thirty-four-year-old widow he left behind, the two fatherless children—one celebrating his third birthday the day of his father's funeral, the other just six years old—made these ceremonies extremely wrenching for many witnesses. Letter writers commented on Mrs. Kennedy's extraordinary stoicism throughout her husband's funeral rites, admiring in her a strength they sought and felt they lacked. Her dignity saved the country, some wrote, from the degradation the assassination visited upon it. Her composure and that of her children amazed those who found it difficult to collect themselves. Many described being deeply moved by Caroline Kennedy's visit to the Rotunda Sunday afternoon with her mother when the President's daughter knelt by her father's casket and placed her hand under the flag that covered it. They were likewise touched by John F. Kennedy Jr.'s salute to his father, as the President's body was removed from St. Matthew's while a band played "Hail to the Chief" for the thirty-fifth President for the last time. During the burial rite in Arlington Cemetery, Mrs. Kennedy and the Kennedy brothers lit the eternal flame that would mark the President's grave. The graveside ceremony riveted the nation, carving deep memories into the consciousness of millions who watched these events on television. Some later gathered in their own communities and places of worship to memorialize the President.

In their letters to Mrs. Kennedy, a few families and neighbors of soldiers who participated in Kennedy's state funeral described their special pride in being part of the solemn and historic occasion. Ironically, one of JFK's most vexing problems as a candidate—his Roman Catholicism—became a virtue in his death as Americans of all faiths observed the rituals of his church, admiring their sacred splendor. "I have been taught all my life," one young North Carolinian later wrote to Mrs. Kennedy, "that Catholics were 'doomed.' I felt we needed a Christian leader. I realized by watching the funeral over television that our church and people who have judged without really knowing are very wrong. I want you to know that was all I had against him and now that's gone."

NOVEMBER 25, 1963

Dear Mrs. Kennedy:

Yesterday my wife and I went to Washington to pay our respects to the President. We didn't know what else to do. We stood on Pennsylvania Avenue bathed in sunshine and trying to believe what was happening. We couldn't do it. It was still unimaginable that a man so much alive could be being borne on a caisson over the same route that he had traveled so triumphantly for his inauguration only two years, ten months, and two days before. The drum beats were so slow, the honor guard so stiff, the change so obvious; it was not real. My mind went back to the 1960 election and I remembered with pleasure how unpopular I was, as one of the four registered Democrats in the town of Essex, Massachusetts, when I went to the polls wearing my oversize Kennedy button and voted for the first time. Among many other things, I will always associate him with the first election in which I was old enough to vote—a moment I had anticipated eagerly for a long, long time. Thinking these thoughts and watching the caisson roll by produced emotions in me that I never knew I had. I remembered him saying in the 1960 campaign: ". . . The American Presidency demands that the President place himself in the thick of the fight, that he care passionately about the fate of the people he leads, that he be willing to serve them at the risk of incurring their momentary displeasure." It was a Jeffersonian remark. And it was what the President was doing when he incurred the momentary displeasure of a madman with a mail-order rifle.

For us, the President's otherwise meaningless death has filled us with a sense of duty—a feeling that when the standard-bearer falls, it behooves everyone to help get the flag up off the ground. Perhaps this is what his death will ultimately prove: that good men still die for their countries, and that we must learn from their deaths to save other good men from the same fate. "Ask not what your country can do for you—ask what you can do for your country." They are good words in a fine tradition. Franklin D. Roosevelt spoke in the same tradition in his third inaugural address, and his words would have made a fitting scripture lesson in the funeral today:

> It is not enough to clothe and feed the body of the Nation,
> and instruct and inform its mind. For there is also the spirit.
> That spirit speaks to us in our daily lives in ways often unno-
> ticed, because they seem so obvious. It speaks to us here in the

capital of the Nation. It speaks to us through the processes of
governing in the sovereignties of forty-eight states. It speaks
to us in our counties, in our cities, in our towns, and in our
villages. It speaks to us from the other nations of the hemi-
sphere, and from those across the seas—the enslaved, as well
as the free. Sometimes we fail to hear or heed these voices of
freedom because to us the privilege of our freedom is such an
old, old story.

A gruesome new chapter in the old, old story was written on November 22. Much of the future depends on what voices we hear and how we
choose to hear them. One of the voices we should have listened to more
was marked with a broad Boston accent that lent itself to frequent parody
rather than rapt attention. Perhaps if we all listen very hard, we can still
hear it for awhile.

Blessings on him, and on you and on your children.

Sincerely,

Thomas N. Bethell

SIGMA PHI EPSILON
LEXINGTON, VIRGINIA
MONDAY
25 NOVEMBER 1963

Dear Mrs. Kennedy,

I know that you will never read this letter personally, but I feel
that I must write it anyway. Countless thousands of letters will
doubtless be pouring into Washington in the next week expressing
the nation's sympathy and condolences, so this will constitute but an
insignificant fraction of this country's conveyance of an inexpressible
feeling of loss.

I am a student at Washington and Lee University, which has a history almost as long as the United States. Two of the university's past

presidents, George Washington and General Robert E. Lee, left in the school an indelible tradition of respect and of honor—respect for men who have contributed to the development of this country and the American way of life, and honor through an unwavering application of a special brand of honesty to one's own life, coupled with a genuine desire to find the same quality in others.

Having had ingrained in me such a background of personal conduct and reverence for the men who shape American history, I came to Washington to pay my respects to the President this past week-end. Though I have been in Washington several times before, never had I seen such a soletude and feeling of extensive sadness. When the President's caisson passed before me, there was not a sound to be heard except for the drums and the horses' hoofs. The smallest children assumed a composure of sobriety I have never before seen in the faces of children, and which I pray I may never have to see again. When I arrived at the Capitol, it seemed as if half the world were waiting in front to pass before their President. When the portable radios in the crowd announced that the President's asassin had died, no one appeared either happy or unhappy— they remained pensive, for they had thoughts of only one man that afternoon, and he wasn't in Dallas.

When I saw you and little John and Caroline emerge from the Capitol, I realized that never, Mrs. Kennedy, have I seen a person display such strength and courage as you did that afternoon. After returning to school, I watched television until well into the morning as thousands of Americans filed past their President, and I saw the same looks of disbelieving grief that I had seen in person that afternoon. I said a prayer before going to bed that night, Mrs. Kennedy. That's something I haven't done for a long, long time. And I can't help wondering whether if more Americans had said theirs before, I wouldn't have had to then...

Most sincerely yours,
Jim Legg, Jr.

ANNAPOLIS JUNCTION,
MARYLAND
APRIL 22, 1964

Dear Mrs. Kennedy:

I dislike reminding you of the death of your fine husband but there is one phase of the ceremonies and events that I was part of and I have never seen it mentioned in any newspaper or magazine articles or heard it mentioned over radio or TV, and I have read and listened to everything I could concerning the last days in Washington.

My wife and I waited on the north side of Constitution Ave. at the Capitol until the caisson & the procession moved past & turned towards the East front of the Capitol. We were then told the line would form on East Capitol Street. We moved as fast as we could and we still had to go all the way to 9th St. N.E. before getting in the line on East Capitol St. I have read and heard that the line was six abreast. It was at least twelve abreast and sometimes more depending on the width of the sidewalk. The people in line were friendly and respectful. We talked to many from as far North as Maine and as far South as Atlanta, Ga. Most came by rail, bus & automobile. We talked to five young men from Atlanta who had come by plane at a cost of over $100 each for fare alone. Many could not wait to get into the Capitol as they had reservations to return home. Some, after waiting many hours had to leave to catch their transportation home so as to get to work on Monday. All were deeply touched. All of those I talked with said they had made up their minds to vote for Mr. Kennedy for President in 1964 although many had not voted for him the first time in 1960.

Now, the most impressive thing that happened as far as I know has not been commented on by either the press, radio or TV. It is as follows:

A group of people would start singing and when that group had finished a group in the block ahead or behind would either pick up the same refrain or some other one. When that was over a group in my block or several blocks away would sing something else. The songs we sang were: God Bless America—My Country Tis of Thee—America the Beautiful—The Star Spangled Banner—Nearer My God to Thee—Onward Christian Soldiers. There were probably other songs in blocks so far removed from me that I couldn't hear them.

It was wonderful to be a part of such a spontaneous demonstration of love & respect for our fallen leader. I do not know why this has not been reported unless it was because all phases of news were concentrated in the Capitol and it was a case of not being "able to see the woods for the forest." I do think that this should be recorded in history and hope you will see that it is. It was cold & windy on East Capitol Street and though people were in line for hours no one complained. Some people were old & crippled and when they got to the Capitol steps they could hardly climb them. I saw a man close to me who carried a three or four year old child on his shoulders for the whole of twelve hours.

From where we were at 9th and East Capitol Streets it took 12 hours to get to the Capitol & through the Rotunda. However, I'm sure the line went out to 17th St. or beyond and then backed up the street one block north of East Capitol. When we came out of the Capitol the line was still forming to go out this street all the way to 17th and then back again on East Capitol towards the Capitol. I think that the number of people who took part in this demonstration of respect & sympathy, on this memorable Sunday was far in excess of the number mentioned in the news, especially considering those that had to leave for home before they go to the Capitol.

I hope I have not bored you with a recount of the happenings on this sad day but I only want history to truly record the way the people felt and acted.

Sincerely,
Arch C. Keegin

GARDEN CITY, MICHIGAN
NOV. 25, 1963

Dear Mrs Kennedy

In your deepest moments of sorrow I am moved to write this letter. It is a few minutes to 2:00 A.M. (Detroit time) I am watching the multitude of people paying homage and saying their last goodby's to a great great man. The lines are now five abrest and three miles long, the reporter just said. If a mans greatness and love of man can ever be seen it is seen in the faces of the people who come to pay their respects at a moment like this. I can see sadness, love, sorrow, disbelief, tenderness, and tears in faces of men, women and children of every creed and color.

The respect that is shown to your husband is one that would warm your heart. You can see Catholics, cross themselves and pray, women carring rosaries and praying. You can see old people bow their heads in reverence to a young man, and you see the young people, the future Americans even the young men in the service of their country stop at the casket and give a proud salute to their President.

Mrs. Kennedy, you have been very brave at a time like this. I hope and pray with all my heart and soul that God will give you the courage you will need in the future. I know the prayers of all the people are with you and your children just as mine are now. God Bless you and give you comfort.

Yours with eyes to the future,
Mrs. William Tomashek

....❧

FRANKLIN, KENTUCKY
1-13-64

Dear Mrs. Kennedy,

I would like to take this opportunity to express our deepest sympathy on behalf of my family. We loved the President as if he were a member of our own family. He was a wonderful young man and we were fortunate to have had him as our leader.

When we heard the news we were deeply shocked along with the rest of the nation. We hardly left the television during the whole thing and we thought you held up wonderfully. We admire you for your great courage.

Our son, who is presently stationed at Fort Myer in Arlington rode in the President's funeral. His name is Sp. 4. Charles B. Wade and he rode the horse just in front of the President's body. He said there weren't many women who could hold up and be as wonderful as you were.

With deepest sympathy for you and your children we say God bless you always.

Sincerely yours,

Mr. and Mrs. Aubrey Wade and family

Charles B. Wade is the third rider to the rear, closest to President Kennedy's casket.

Dear Mrs. Kennedy,

I am enclosing a typewritten copy of a small town newspaper article containing a letter which I thought perhaps you and, or your children at a later date may be interested in. As you will note the letter, as published, was written by a member of our armed forces whose duty it was to help honor your late husband, our great President, at his funeral in Washington, D.C.

I personally know some of Mr. Hendrickson's relatives but have not informed them of my intentions of forwarding this to your attention. I guess, however, that they would be deeply honored.

All of us in our family were deeply grieved at the death of your late husband and extend to you and your family our deepest belated sympathy.

Yours truly, Melvin Parrish

ATHENS FREE PRESS of Athens, Illinois

Issue of December 6, 1963.

Mike Hendrikson, of Athens, a son of Mr. and Mrs. Eldon Hendrikson of Enumclaw, Washington, formerly of Athens, had a part in the Army's part of the late President's funeral and burial rites. Mike is a graduate of the Athens High School and he joined the Service soon after graduation. He is stationed at Ft. Meyer, Arlington, Virginia; and part of his duties consist of being a part of or cooperating in the meeting of foreign dignitaries visiting this country and taking part in other state functions.

We are indebted to his sister, Mrs. Robert Chastain of R.R. 5, Springfield, Illinois for the use of this letter.

Dear Sis and all:

The last time I saw Kennedy close enough to touch was on Veteran's Day (eleven days prior to the assassination). He laid a wreath on the Tomb of the Unknown Soldier.

On the twenty-second of this month, we were out in the cemetery having funerals. We got back at 2:00 p.m. EST. While we were getting off the bus I noticed there were about 25 or 30 guys standing out in the back, gathered around a radio. I knew something was wrong, so I went over and asked what was going on. One guy turned around and said, "They say our boy Johnnie has been shot, but I didn't believe it. I ran in and turned my weapon in and went up stairs, got my radio, sat down, and listened. The word came down from Headquarters that we would be on a 72 hour ceremonial alert. So we had to get our dress blues ready and stand by, ready to move.

Saturday, November 23, the whole Battalion went out on parade field and the Battalion Commander read the Presidential Death Order to us—It's really odd, but it was pouring down rain out there and no one said a word about it, no one seemed to even notice the rain or even the fact that we were all getting soaked. We were not wearing raincoats either. That was about all that happened Saturday except that we stayed around the company on alert.

Then Sunday, November 24th, we had reveille at 6:00 A.M. Then we got orders to move down to the Capitol Building where they were going to move the late President's body for the people of his nation. I was standing right at the top of the steps where they took the body. C.B.S. news cameras and radio were directly behind me. I could listen and watch at the same time. As the casket came up the steps, Mrs. Kennedy and her two children immediately followed it. Then came the heads of our government. As Mrs. Kennedy reached the top steps she then turned and looked over my shoulder to the cameras. She hesitated on the top step, then gave a faint smile, as to say, "Thank you Americans for the feeling and understanding." This may sound silly but I got a lump in my throat. After the ceremony we came back.

But this is really a touching part of the whole thing. I was really cold, but people from where I don't know, came into the rotunda all night until 10:00 A.M., Monday. But as cold as it was they waited in lines nine and ten hours, waiting, shivering, but devoted to seeing their great leader one more time. Also people by the thousands slept in lines, in sleeping bags all night, along the route to Arlington National Cemetery where he would travel the next day at 1:00 p.m. We waited around before we moved into position. Meanwhile I went up to the casket in the Rotunda.

After the ceremony there we left and went to eat, then to the Cemetery. I was posted right below the grave site. We waited down there for about two hours. I think you know that in all ceremonies we stand at either parade rest or at attention. We were standing at parade rest. When the colors and the casket came by me I was so numb I couldn't complete a present arms! (I couldn't move).

Oh I wish you could have seen the funeral and heard it also. We could hear them coming across Lincoln Memorial Bridge. Those drums were beating a death March. They were really beautiful.

I guess that's about all, except for the feeling of the people. Honestly it was enough to bring tears to your eyes. We were watching the people. They were walking along and then suddenly they just fell on their knees and prayed right on the spot, in the streets, along the sidewalks, in parks, everywhere.

People walked around in silence. The city (DC) had normal traffic, but there were no sounds coming from anywhere. It was like a silent movie. It hit all of us here in the Old Guard very hard indeed, but some would joke and laugh to hide their sorrow, while most of us just sat around.

I remember on Veteran's Day, President Kennedy laid a

wreath on the Tomb, and he had Johnnie, Jr. with him. As they rode by us Johnnie, Jr. was crawling on his dad's lap, saluting us as they passed. Seemed like any other father and son. You should see Jacqueline now, she looks as if she has aged ten years.

I am making a scrap book of it all to keep. This was one of the most proud moments of my life. It's history, and we all played a part in it. You asked if I could be seen on T.V. Yes, I didn't know it but while they were taking the casket into and out of the Capitol Building they had cameras directly on me while the U.S. Marine Band played "Hail to the Chief"— which as for a couple of minutes. When I got back to the C.O. some of the N.C.O.s who were watching it on T.V., got after me for moving while the cameras were on me. But we had only been standing there about three and one-half hours at attention.

You should have seen the people. It finally totaled well over a million. The streets were packed thirty or forty deep, and across the bridge they were twenty deep, this was on both sides. Then at the cemetery there were ropes around the grave site about 200 feet all the way around. They were packed all the way about a thousand deep or so.

Countries which I have never heard of before had sent representatives to the funeral. A few of us were talking to the chauffer of the President's Bubble-top car. He said the Robert Kennedy, Mrs. Kennedy, and Edward Kennedy just didn't seem to realize what had happened. He said that they just didn't say anything but just sat and stared at nothing.

Well, it is one of the most tragic things that has ever happened in this country, but we, like time can't stop. One thing before I close while all of this was happening, I wanted to be with my family. It was as though it was one of us. But God knows how hard it hit the people of Washington, D.C.

Would you please keep this letter for me—it's like a diary
for me. Then I can remember how I felt and what I did.
Merry Christmas and God Bless you all.

Love,
Mike

.... ❊

GEORGE C. LODGE
BEVERLY MASSACHUSETTS
NOV 26, 1963

Dear Jackie:

I know that you are keenly aware of the great shock and sadness which
the President's death brought to this nation; it can only be a pallid reflection of your own. I happened to have been in Washington on that dreadful Friday and wandered aimlessly through the silent streets for hours,
part of the tearful, desperate crowd, bewildered by the news, angered,
confused, and, most of all, terribly sad and not a little frightened that our
nation should have been so senselessly and cruelly deprived of a great
leader, which it badly needed.

But what you cannot know so well is the incalculable strength which
came to the millions who watched the fortitude of you and your family during the funeral ceremonies. I doubt that ever in history has a nation witnessed
such a thoroughly inspiring sequence of events. While they were enriched by
the trappings of history and tradition and given meaning by the glory of religion, they were unique in their grandeur by virtue of the very special nobility
given to them by you and your children. The great sense of loss cannot be
blunted, but what was blind misery and helplessness has become converted
into that spirit of confidence, dedication to high purpose and victory which is
so beautifully stated in the President's Inaugural Address.

Nancy joins me in sending you and your children our love and
boundless sympathy.

George

PHILADELPHIA, PA.
DEC. 1, 1963

Dear Mrs. Kennedy:

Words cannot express our feelings at this point for our President John F. Kennedy. He was loved by many, hated by some, feared by some, but respected by all. He had many ideas; but with a single thought; to move America forward and make America one unit of people.

We are a Negro family of three (3). I am 31 years of age. My wife is a nurse and we have a three (3) year old son. Last Friday my son Bruce told me (when I came home from work) "Daddy, the President is dead." I knew that he did not reallize what this meant. But I did and the whole world knew. As big and hard as I am, tears over-flowed and I broke down. I knew that I lost a President and a good friend.

He was a great man and a great President. How-ever, behind every great man there is a great woman. My hat is off to you. You were with him throughout his years on Capital Hill right up to that fateful hour. You have set a standard that no woman could possibly equal. The whole world is proud of you and what you stand for.

I have seen funerals for Presidents, Heads of State, and digniteries, but John F. Kennedy's funeral was the most fantastic and the most beautiful I have ever witnessed.

President Kennedy was a God sent man, sent for a purpose and he was attempting to complete his task when he was cut down. Even in death his mission is being accomplished.

John F. Kennedy will forever be a burning memory in my heart. May God bless you and may He always keep you as strong as you are.

Sincerely Yours
Mr. & Mrs. Hugh B. Robinson, Jr.

P.S. I am proud to be an American and to have lived under your husband's administration.

Dear Mrs. Kennedy:

I wanted to come to the President's funeral so bad, I felt as if I had lost a brother of my own, and I still do, I intend to see his grave one of these day's on my vacation, I feel so sorry for John Jr. and Caroline, for I know they miss their father, I really loved the man myself, I feel sorry for you too, Please don't let Robert or his brother get south of the Mason Dixon for I am afraid for their lives too.

I HOpe you will be able to live your loss down, I don't think the country will ever be able to.

Your Friend
E. G. Martin

ST. LOUIS COUNTY
NOV 28/1963
MRS. J KENNEDY AND HER TWO SWEET CHILDREN

Dear Sis

This is a Letter with Grief and Sory From My Bottom of My Heart and Sympathy to you Mrs J Kennedy and Children and the Grief and Sorry for Your Dear Husband Es [Ex] Presedent Mr John Kennedy That Shall Never Be Fargotton By Anbody are the Whole World as the Spirit of All American People of the United States of Amerca

The Spirit That God Gave Es [Ex] Present Mr. John Kennedy of the United States of Amerca. He whas Loved For His Truthful Speres to The Whole World and He tried To Be Friendly to Everdody Poor Peopel to Rich Peopel and Foes and Enemies and He Spoke the Truth from His Bottom of His heat. and He Never Be forgoton By the Whole World. God Have Mercy on him. . .

I am Happy to write a few lines. I Hope this Letter will Give and your Children a Spirit and We Loved You All God Bless you and the Boy and Girl The world Knows and Sees How You Happy Children acted all through the Performent—What Touges the People Heart—When You Mrs Kennedy and Kneeling at the Coffin of your Husband Mr J Kennedy and your Daughter Kneeling at your side. I notice and the World Sees Your Daughter Touges with Her Gentle Hand the United States of Amerca Flag that Show that She She Loved Her Farther and Bid Him Goodby, and the Boy How He acted Like a Little Man Mrs Jacklin Kennedy and Your Children I Hope you all Be Happy Good Health and Good Luck. And there Is one that We Never Fargit By the World the one We Loved So Well. With His True Stories that Has toughes the Whole World Mr Es [Ex] Preseden John Kennedy. We Clased the Letter with the Best of Love to Mr Jacklin Kennedy and Her two Sweet Children. God Bless You All

We Closed this Letter With Love and Good Spirichal to Mrs J Kennedy and Children—Mrs J Kennedy I thank you for the Letter I got about 6 month ago When I wrote to you. God Bless you all

> Yours Rests
> Fred Buerman
> and My Daughter
> Mrs Gertude Guidry
> And My two Great Great Children
> My two Great Great Grand Daughters
> there names are Mis Debby N. Krambecht
> and Mis Cheryl M. Krambecht
> Excuse for Writing theis
> My name is Fred Buerman
> I am 92 Years old. Had My Legs take off
> 10 years ago
> I Writing In a Wheel Chair
> I Was Born Sept 25/1871
> God Bless all of You

Dear Mrs. Kennedy:

It is just about two months since your terrible tragedy occurred and we are a little late in expressing our sorrow to you, but it is heartfelt , nonetheless. We would like to tell you that on the day of your husband's funeral, our synagogue conducted a most moving service with the Rabbi and the Cantor chanting innumerable Prayers and blessings.

In the Jewish religion, when a father dies, the son must stand and say "Kaddish"—a most meaningful prayer and only permitted by the sons. In our synagogue, West Oak Lane Jewish Community Center, our Rabbi told us that we had all just lost our father, and the entire congregation stood up to say Kaddish.

At the conclusion of the service, the Rabbi and Cantor marched down the aisle Chanting prayers as if following the casket.

Mrs. Kennedy—these words may be added to the tons of condolences you have received and may sound quite repetitious, but none can be more sincere.

May God bless you and your family and may you all live to ripe old ages and have many great joys through one another.

Most sincerely,

Pauline and Sol K. Spector and family

DALLAS, TEXAS
JANUARY 16, 1964

Dear Mrs. Kennedy:

I have put off writing you as long as I can. I have wanted to write for several weeks but hesitated, as I felt the things I would want to say would be like opening an old wound and would hurt you too much. After seeing and hearing you on T.V. last night, and you stated that you only read them as you felt you could bear it, I decided not to delay writing another day.

First, I'd like to extend my deepest and sincerest sympathy for you and your adorable children. Words are really inadequate. I also wish there were ways to express our sympathy to the nation in their loss and also to the world. Nothing to compare will ever effect the globe like this sad tragedy.

That rainy Friday morning I sent a note for my 14 year old son to be excused from school at 10:30 a.m. so that he could come to the parade to see the President. We were so disappointed it was raining. He came downtown and met me and the sun came out brilliantly. We stood on the corner by the fameous Neiman-Marcus Store and waved to you as you drove by. You both looked so happy, you so pretty and he so handsome as the sun glistened on your bright smiles. The crowd seemed delirious with excitement as thousands on the streets and hundreds leaned out windows, in their bright colors, cheering. My son remarked that "Jackie was prettier than her pictures"!—I overheard a little colored boy say with animated enthusiasm, "I was so close to him I could look right down in on his eyes!" My son caught a bus and started back to school.

I thought, "What a moment to remember!" As I walked toward a downtown store for lunch, I was thinking about seeing all of this to night on T.V. and hearing what the President would say to Dallas. Then, two minutes later—Eternity!

You would have known how shocked and stunned Dallas was if you could have seen the people. They were weaping unashamedly both men and women, wringing their hands, some praying aloud and mostly wandering around in a daze, as if they must go somewhere and do something, but did not know where to go or what to do!

I don't remember how or when I found my way back to my office. Surely there could not have been more despair, nor could it have been more earth shattering if it had been the end of the world!

I personally was in such state of shock, the rest of that day is still almost a blank. I think the thing that brought me back to reality, was the T.V. showing how you exemplified such gallant bravery in everything you did. I shall never forget your standing there for Pres. Johnson to be sworn into office. Truly your actions were queenly, which your husband would have been justly proud.

Of course those next 3 days are to go down in history and generations for years to come will know of those 4 saddest days in 1963, starting here in Dallas.

The two things most heartbreaking and indelible in my mind, was seeing little Carolyn, whom he loved so dearly and we all felt like we knew her well, standing so ladylike and so like a little princess, beside you at the Services in the Rotunda. And never, never will I, or the nation forget that precious little boy grasping the U.S. Flag and running down the steps to take it to "my Daddy!"

If any one moment in time could be recalled, I'm sure that moment, November 22, 1963 would be the one. Be assured many, many more feel deeply concerning this even though they have not written you.

To prove we have no hate or bigotry in our home, my husband and I are Republicans, and our 2 unmarried older sons proclaimed JFK as the pattern of the young and future leaders of America. The older one, at age 23, has already served in the Navy and is in 3rd year college, the other, in college also and eagerly awaiting he service with the Peace Corp, is already signed in and has his official Peace Corp number. We are convinced the young people of our nation has been as hard hit as the adults.

If this will help any, I am a christian and am known to be a truthful woman and would be afraid to write this if it were not the exact words and truth. The day the President was killed, I vowed to God, in my prayers that I would gladly have died, if it could have brought the President back to us.

We pray that as the flame burns magnificently on, that it will in some way burn out the sad and horrible memories you must have of Dallas.

Affectionately and sympathetically,
Mrs. Carroll A. Geist

<u>STEPHEN J. HANRAHAN</u>   <u>DECEMBER 2ND 1963</u>
        85255   (DATE)
TO   <u>MRS JACQUELINE KENNEDY</u> <u>1600 PENNSYLVANIA AVE.</u>
                                  <u>WASHINGTON, D.C.</u>
        (NAME)   (ADDRESS)

Dear Mrs Kennedy:

   I wish to extend to you and the children my condolences. Children increase the cares of life but they do help to mitigate the remembrance of death.

   We are told that a good key is necessary to enter paradise. The President, following the guidelines of his church possessed this key. Heaven it seems, calls its favorites early.

   In the President, I felt that I had known a whole man. It is a rare experience but always an illuminating and enobling one. It costs so much to be a full human being that there are very few who have the enlightenment, or the courage to pay the price.

   The lights of the prison have gone out now. In this, the quiet time, I can't help but feel, that my thoughts and the thoughts of my countrymen will ever reach out to that light on an Arlington hillside for sustenance.

   How far that little light throws his beam.

   Sincerely

   Stephen J. Hanrahan 85255

   Federal Penitentiary

   Atlanta, Georgia

ROCKVILLE, CONN.
NOVEMBER 28, 1963.

My dear Mrs. Kennedy:

   I could not let this week come full circle without expressing deepest sympathy in the great loss of your beloved husband and our beloved President.

Truly, the whole world mourns his passing, and one can only hope that, in some way, your own heartache may be mysteriously lightened by the commiseration of so many who share so strong a sense of personal loss.

The impact is everywhere. You drive down country lanes, on off the beaten track roads, and outside of house after house there is the flag of our country fluttering at half mast. Fellow citizens mourning outside their homes as well as in their hearts. It tugs at the throat, but you go on only to have the next friend encountered inquire: Have you noticed that the ache keeps coming back again and again?

It does, of course, but it is a precious thing this grief he leaves us, for the sorrow now would not be so deep if the joy and pride in him and you had not been so high.

All that he represented, all that he was, his children, his First Lady, and our First Lady ever to be, remains enshrined in our hearts, our memory, our love.

Sincerely,
Leonard F. Rock

POLITICS, SOCIETY, AND PRESIDENT, 1963

"He Was Born Holding a Flag"

Columbia Mo
Dec 3 - 19-6 3

Mrs Jacqueline Kennedy
You have my greadis Sampthia
Just a week be fore President
Kennedy died he sent me
Certificat Memory Armed Forces
of LafeE.Stone My Son he was Kill in
July 27 - 1953.
Then i a little boy to get Kill on
a horse at 12 years old.
I Know part what you are going throw
with. Know one nows onely one asto
go throw with it.
President Kennedy was like by every
on i hard Talk, he seams just to see
him on T.V. our picture in paper
such home Nice pearson.
I work as maid at Dorn E Universty
of Columbia Mo, he was to Come to
Dorm i was work at on See ing
him Plawing
This Country has lost a good President
He look so Kind —
I lived by self - work as maid. in
a Partment by self So i Thought i write
tell you he was well like by me—over

resident Kennedy's assassination led to much soul searching in the nation. For weeks, months—even decades—
exhaustive analysis and debate about JFK's death and its
meaning filled the print media and airways. A variety of politicians, commentators, insiders, and intellectuals sought to uncover the reasons for the
assassination, to measure its impact on the country and to assess Kennedy's
legacy. The latter endeavor proved especially problematic given the brevity
of his administration, its sudden violent conclusion, and the persistent grief
that soon created a mythic figure of a martyred President. "It will not be
easy for historians to compare John Kennedy with his predecessors and suc-
cessors," his brilliant Special Counsel and speechwriter Theodore Sorensen
predicted two years after the assassination. But, he ventured, "people will
remember not only what he did but what he stood for." That fact, he hoped,
would "help the historians assess his Presidency."

Ironically, however, some of the most compelling commentary about the
assassination's impact on Americans never reached the public. In the privacy
of their homes, dorm rooms, barracks, and offices, citizens themselves—
"average" Americans all—gathered pen and paper to record in letters to
Jacqueline Kennedy their own views about the President. In so doing, they
revealed much about how they saw the nation. Working men and women,
those who lived in poverty, immigrants, those suffering from physical illness
and disability, minorities who especially believed they had a champion in
Kennedy, and Catholics who took singular pride in the election of the first

President of their faith described very openly why they grieved his death. Their letters vividly portray thoughtful citizens seeking solace, wisdom, and understanding.

Some blamed the assassination on what they saw as the country's political, social, and moral failings. A few were convinced that through John F. Kennedy's brutal death a vengeful God had punished a sinful nation. Others pointed to their own perceived personal failings as disinterested citizens. Many more described a political climate that they believed permitted political excess, extremism, and hatred to flourish. References to Kennedy's Catholicism—and the open opposition it inspired during the 1960 election—abound in the condolence letters, especially among those who believed that bigotry had contributed to a toxic climate that might have encouraged an assassin. "It still riles me to think", a citizen confessed, "that the only Catholic President we ever had should be taken from us in this terrible manner." She recalled her mother's family's attempt to build a Catholic church in New Jersey: "When the church was being built the Ku Klux Klan would tear down whatever was put up that day and the Catholics would have to start all over again but they stuck to it ... If we had a man like Mr. Kennedy in office then perhaps this wouldn't have happened." Others located the fault for the President's assassination squarely in rigid racial prejudices harbored by the American public. "We may not realize that fault now," wrote one Pennsylvanian, "but in time, history will write its own story on blood filled pages."

Those who took the measure of American society in 1963 commented most frequently on racial unrest. In so doing, the letters offer striking evidence of how deeply embedded segregation remained in the nation at the time of Kennedy's assassination. They also reveal the tensions that challenges to segregation had heightened—especially among those who had watched the civil rights movement only with discomfort. "We did not agree with your husband," wrote not a few Americans, from every part of the country, who took the time to pen a condolence letter to Mrs. Kennedy. Such letters came from every part of the country, but often the source of the distress was a civil rights policy some Americans viewed as far too aggressive. A Louisianan who admitted that she "raged against the President when he was in office" was joined in that sentiment by a native of Mississippi who implored Jacqueline Kennedy to do whatever she could now to prevent passage of her husband's civil rights legislation. A rare letter from a white supremacist depicted President Kennedy as a "wayward son" who supported "black-white mixture" because he had been hoodwinked by liberals, Communists, and Zionists. Many white Southerners wrote to say they knew Kennedy was right in emphasizing racial equality, though they "lacked the courage" to say so in a climate they believed imperiled the life and limb of anyone sympathetic to desegregation. Despite the lack of any evidence tying Lee Harvey Oswald to white supremacist organizations or anti–civil rights sentiment, some letter writers—black and white—expressed the firm conviction that Kennedy's death marked another regressive milestone in the long struggle to achieve racial equality.

Such fears were the flip side of another view of the nation expressed very often in the condolence letters—the powerful mood of optimism and idealism, cutting across generations, that many Americans believed President Kennedy had created. Young and old, World War II veterans and their children, rich and poor, dyed-in-the-wool Democrats and political atheists alike describe being inspired by the rhetoric and aspirations of John F. Kennedy. On the night of the assassination, one New Yorker recounted her love of country and her conviction that Kennedy exemplified its strengths. "The day he was born, he never belong to his mamma or daddy because God put a Star on him. he belonged to his country," she wrote. "He was born holding a flag." Many saw in Kennedy their own dreams for the nation. They lauded JFK as a leader who sought to overcome the country's divisions and realize its promise. For some, the assassination leveled a harshly felt blow against these soaring ambitions. Virtually all letter writers who mentioned Lyndon Johnson offered their support, high regard, best wishes, and praise for the new President even as they mourned his predecessor. But one can also feel an effort to divine a way forward for the country in the face of a devastating disappointment.

. . .

1963 had already been a very tumultuous year before the President took his fateful trip to Dallas. Throughout the spring, summer, and fall, the civil rights movement was widening its challenge to segregation, pressing not only the case of racial equality but the Kennedy administration's failure to act with clear affirmation, commitment, and speed. During the 1960 election, Kennedy had won over African Americans with his promise to redress discrimination, including a vow to end housing discrimination with "the stroke of a pen." He earned admiration, too, for his actions when Martin Luther King Jr. was jailed during a sit-in and sentenced to four months of hard labor in a Georgia penitentiary. Political calculation figured in the equation, but JFK's sympathetic phone call to Coretta Scott King and Robert Kennedy's pressure on the judge to reduce the sentence convinced some African Americans that Kennedy held the best promise for their race. In the cliff-hanger 1960 election, over 70 percent of African Americans voted for Kennedy.

But after his election, Kennedy's inattention and caution bitterly disappointed many civil rights activists. The federal government's failure to protect those engaged in freedom rides and voter registration drives in 1961 and 1962 had cost lives. Insisting that the federal government had no jurisdiction to intervene, the Kennedy administration left state and local officials in charge even when the latter refused to prevent (at best) violent attacks on civil rights workers. The resulting bloodshed and ongoing threat of violence badly damaged the civil rights movement's faith in Kennedy. Three years into the administration, many still awaited civil rights legislation.

In April and May, the Southern Christian Leadership Conference embarked upon a bold campaign to speed up desegregation and move the Kennedy administration. They brought the movement into one of the most segregated cities in the country—Birmingham, Alabama. Despite their blatant unconstitutionality, local ordinances continued to specify rigid segregation of the races, with separate drinking fountains, restaurants, schools, and facilities of many kinds the norm. *New York Times* correspondent Harrison Salisbury described the city as a place where "whites and blacks still walk the same streets. But the streets, the water supply and the sewer system are about the only public facilities they share." Here Jim Crow was enforced by "the whip, the razor, the gun, the bomb, the torch, the club, the knife, the mob, the police and many branches of the state's apparatus," he asserted. To confront the degrading conditions in Birmingham would be to dramatize the political, financial, moral, and human costs of the Kennedy administration's reliance on legal enforcement of desegregation orders by local officials.

As expected, it did not take long for events to escalate. In April, sit-ins at Birmingham lunch counters, boycotts, and marches led to the arrests of civil rights demonstrators, including Martin Luther King Jr. From his jail cell, King wrote his eloquent "Letter from Birmingham Jail," explaining the urgency, moral purpose, and philosophical underpinnings of the action and of civil disobedience. May brought the "children's crusade"—a series of marches that attracted national attention and outrage when city officials turned snarling police dogs and fire hoses capable of ripping the bark off of trees, onto demonstrators, including six-year-old schoolchildren. The scene, Kennedy said, made him "sick." By May 10, the SCLC campaign produced an agreement, brokered with the assistance of the Justice Department, that would mandate the dismantling of some of the most egregious Jim Crow practices in the city.

With summer came further highly publicized confrontations, this time with more aggressive intervention by President Kennedy. Governor George Wallace's attempt to prevent the court-ordered desegregation of the University of Alabama produced an open test of wills with the Kennedy administration. In the end, Wallace relented but not without offering a ringing defense of states rights even as he faced a general from the Alabama National Guard, now federalized, and Justice Department officials. The very evening of the showdown with Wallace, Kennedy delivered a forceful civil rights speech on national television announcing his intention to introduce new civil rights legislation. Although other American presidents had previously taken occasional affirmative steps to address institutionalized racism—Harry Truman's efforts to desegregate the armed forces being, perhaps, the most notable example—Kennedy's pledge in June of 1963 to pursue sweeping federal civil rights legislation made a powerful impression.

"We are confronted primarily," Kennedy said on the evening of June 11, speaking in part extemporaneously, "with a moral issue. It is as old as the scriptures and is as clear as the American Constitution. The heart of the question is whether all Americans are to be afforded equal rights and equal opportunities, whether we are going to treat our fellow Americans as we want to be treated." Noting that "one hundred years of delay have passed since President Lincoln freed the slaves, yet their heirs, their grandsons are not fully free," Kennedy insisted, "this Nation, for all its hopes and all its boasts, will not be fully free until all its citizens are free." "Now the time has come," Kennedy continued, "for this Nation to fulfill its promise." Events in Birmingham and elsewhere had proved that the "moral crisis" facing the society could not "be quieted by token moves or talk. It is time to act in Congress, in your State and local legislative body and, above all, in all of our daily lives. It is not enough to pin the blame on others, to say this is a problem of one section of the country or another, or deplore the fact

President Kennedy met with leaders of the March on Washington at the close of the march on August 28, 1963. Martin Luther King Jr. is in the front row, on the left. John Lewis is standing behind King and to his right. A. Philip Randolph is standing to the left of the President, with Whitney Young and Floyd McKissick on the far right.

that we face. A great change is at hand, and our task, our obligation, is to make that revolution, that change, peaceful and constructive for all. Those who do nothing are inviting shame as well as violence. Those who act boldly are recognizing right as well as reality."

Although Kennedy worked in the ensuing months of his life toward approval of a civil rights bill, "the fires of frustration and discord" continued to burn. The very evening of JFK's civil rights speech, Mississippi civil rights activist Medgar Evers was shot in the back by a Ku Klux Klansman as Evers returned home from a meeting with lawyers from the NAACP. He died less than an hour later. The March on Washington for Jobs and Freedom in late August, organized by an array of civil rights activists and organizations, dramatized the urgency of the issues and demonstrated to the nation the breadth and depth of commitment. Among the speakers was John Lewis, the young president of the Student Nonviolent Coordinating Committee, who reluctantly stepped back from branding Kennedy's civil rights legislation as "too little and too late." Noting that the "party of Kennedy is also the party of Eastland" (the latter a Mississippi senator outspoken in his support for segregation), he attacked "politicians who build their careers on immoral compromises." And he asked, "Where is our party? Where is the political party that will make it unnecessary to march on Washington?" The peaceful march drew hundreds of thousands and was a powerful testament to the courage and determination of the civil rights movement.

But the violent resistance from some quarters continued. In September the 16th Street Baptist Church in Birmingham, Alabama, was bombed, in an act of terror organized by the Klan that killed four young girls. Although Kennedy faced civil rights leaders who viewed his commitment to the cause as anemic if not woefully deficient for much of his administration, his assassination in November led many Americans to connect his murder to this climate of violence.

It is impossible to quantify the number of condolence letters written to Jacqueline Kennedy by African Americans. Only those who mentioned their race can be identified with certainty, but there are hundreds of such letters in the collection. The fact that so many letter writers mentioned their race is itself revealing. "You and yours have suffered a great loss, by my people and I have to endure even a greater one," read one such message. "You see I am a Negro. For a person to take a stand for what he believes in this day and age, is remarkable. Only a God fearing man would take this stand for the minority." If civil rights activists clearly saw Kennedy's limitations, many letter writers appeared to have been deeply moved nonetheless by his stand against segregation and his initiation of civil rights legislation. "I am one person speaking for all of Harlem, men, women and children," wrote one high school student to Mrs. Kennedy. "This brief, but warm and sympathetic letter is to let you know that Harlem, along with every greatful American shares in your grief." Noting that JFK stood out as one of the "few presidents that ever lifted a finger to help the American Negro," she explained, "that is especially why we are heart broken." This sentiment was echoed in the many eloquent letters African Americans wrote after Kennedy's assassination. Whites, both sympathetic and hostile to Kennedy's position on civil rights, likewise noted the centrality of the issue. ❧

TELEGRAM
LSE310 NSA233
NS JZA022 PD JACKSON MISS 23 700A CST
MRS JOHN F KENNEDY

WHITE HOUSEWASHDC
I EXTEND TO YOU AND YOUR FAMILY MY SINCEREST CONDOLENCE ON THE TRAGIC DEATH OF YOUR HUSBAND. I KNOW WORDS CAN BE OF LITTLE COMFORT NOW FOR I LOST MY HUSBAND ON JUNE 12TH IN THE SAME WAY. THE ENTIRE WORLD SHARES YOUR GREAT LOSS AND SORROW.

MRS. MEDGAR EVERS

WALNUT CREEK, CALIFORNIA
NOV. 27, 1963

Dear Mrs. Kennedy,

There are no words that can adequately express our family's deep sympathy and feeling for you, your children and indeed all the Kennedy Clan.

We are a middle class Negro family and had of course felt after so long that President was like a beacon—a light in the darkness who would indeed be a second Emancipator.

We saw him last year when he came to U.C. My husband a physician even took off. My daughter was inspired by his ideals + goals. She was in the street and close enough she says to see the strength and courage in his deep blue eyes. That he would carry through in his aims.

Now, because of hate and violence he is gone. One who had so much to give. But you and your children can be so proud that he died for his country. If that can bring peace. May God give you the strength and will to go on and live for your children.

If you do ever have time and an extra autographed picture of you two would you please send to my daughter Connie. She is away in Spokane at University. She was so upset she called us.

I think you too inspired the young adults that it is not square to like cultural things. For your excellent example I thank you.

We shall remember you and him in our prayers and Masses. We were so proud of our Catholic President.

> With deepest sympathy
> Cornelia M. Davis

Dear Mrs. Kennedy,

I know you have received quite a few letters in the past week. I pray you take just a moment of your time to read one more.

You won't remember hearing from me, but I wrote to you last October (1962) when my State was getting ready to elect a Gov. and your husband was com'ing to Michigan to make some speaches. I also read somewhere at the same time that he would'nt receive such a warm welcome, and that is the reason I wrote to you and him, asking you to stop by my house for coffee.

At the time that I wrote to you, the President was being bombarded from left and right by everyone and his brother. I told you in the letter I wrote that I had no bone to pick, I was only writ'ing for friendship. I had an answer from your Sectary, saying that you thanked me for my interest in your family. I was so very happy to get that answer.

Mrs. Kennedy, I am colored woman. Your husband made me proud of being colored, by the lov'ing enterest he took in my people, and not only my people, but all the peoples of the world. I wish I had the words, and the knowl-edge of how to put them down on this peice of paper, to tell you what I really feel. I don't even know how to spell good, but I hope that you will understand.

GOD bless you and your children Mrs. Kennedy. I will pray every night that HE will very soon heal the hurt that lie in your heart right now. I will bring this letter to a close, hop'ing that you might read it, although I know that you must have received a million letters by now. This is a terrible time for the whole world, but this, for you, is the worst time of all.

This is what I would like to see erected to the memory of our Presi-dent. It has even been in my dreams.

May the Lov'ing GOD of us all, hold you and your children in HIS hand and carry you though this terrible time.

Sincerley Yours,

Ethel C. Williams

WD 310

NL PD FLINT MICH 23

MRS JOHN F KENNEDY

THE WHITE HOUSE WASHDC

MY HEART GOES OUT TO YOU IN YOUR SORROW BECAUSE IT IS MY
SORROW TOO WHEN THE NEWS FLASHED ACROSS THE TELEVISION
SCREEN OF YOUR HUSBANDS DEATH I CRIED MY HEART OUT MY LITTLE
GIRLS SAID DONT CRY MOMMY BUT I COULDN'T HELP IT HE HAD
GIVEN ME HOPE AND MADE ME BELIEVE IN DEMOCRACY AGAIN BEING
A NEGRO I HAD BECOME DISALLUSIONED BUT TRUE TO MY COUNTRY
WHEN YOUR HUSBAND SPOKE TO THE NATION AND SAID A NEGRO
HAS A RIGHT TO VOTE HE IS A HUMAN BEING HE IS ONLY ASKING FOR
THE RIGHTS THAT THIS COUNTRY WAS FOUNDED ON EQUALITY FOR
ALL MANKIND I STOOD IN MY LIVING ROOM IN UTTER DISBELIEF AND
SAID TO MYSELF WHY A MAN LIKE THIS WITH WEALTH BACKGROUND
LOOKS PERSONALITY RISKS EVERYTHING FOR A NEGRO I SHALL NEVER
FORGET HIM BECAUSE HE WAS A MAN OF COURAGE THERE IS ONLY
ONE OTHER MAN LIKE HIM AND THAT WAS EX GOVERNOR WILLIAMS
UNLESS THE SOUTH STANDS UP FOR WHAT IS RIGHT AND JUST I FEEL
THAT THEY WILL DESTORY US AS A NATION I WILL PRAY FOR YOU AND
YOUR FAMILY THAT GOD WILL GIVE YOU COURAGE SINCERELY YOURS
MRS E SCOTT FLINT MICH

NEW YORK NY

NOV 22 409P EST

MRS JOHN F KENNEDY

THE WHITE HOUSE

MY DEEPEST SYMPATHY TO YOU AND YOUR FAMILY IN THE LOSS OF
A GREAT AND BELOVED AMERICAN FOR WHOM I AMONG MILLIONS
GRIEVE

LANGSTON HUGHES.

SHELBY, N.C.
JAN 17, 64

Dear Mrs Kennedy,

You & your family have our sorrow of the death of your husband not
~~only~~ because he tried to help us as negro but all so he was human but
we feel that Oswald didnot do it it was someone larger than he such as
Mr. Goldwater. We loved your husband because he thought negroes was
Gods love and made us like he did white people and did not make us as
dogs. Mrs Kennedy we are praying for you and your family

Yours truly
Mrs Frank Borders
Shelby, N.C.

WHITE CASTLE, LA

Dear Mrs Kenndy,

I am a color house wife age 49 years of old mother of 10 children 6
Boys 4 girls

I am writting to let you know that my family & I share our heartfelt
sympathy to you & your little ones

I feels so hurted I was left in 1940 with 4 small tots & a baby to Be
borne 3 months later. But you havent only lost a Husband & Father But a
Hero, a man, the only man was Bringing our Race of peoples to the light.
We have lost a Dear friend our hearts [hurt], our homes are Sadden By
the hands of Death I couldent Beleve the News, we cryed, the great man is
sleeping in Jesus.

Mrs. Kennedy you are a Brave Lady. I Pray God through your dark & Sad
hour you finds comfort in him who does all things well. In the great judge-
ment, I'll meet my president. May God Bless you & your little ones always

Daisy H. McKenney

CHATTANOOGA, TENN
DECEMBER 17, 1963

To: Mrs. Jacqueline Kennedy

I just want you to know how much your husband President Kennedy meant to me.

I would like to have one of the True Fact Books that is now being published.

I was born in Clark County, Ga about 1893, and my Father was sold to this Country as a slave. His father was bought by Bill H. Talmadge and later the Allen Talmadge family.

I live in Chattanooga and I have been here since 1922.

My heart and my prayers go out to you and your family in the name of the Lord and your trust in God.

> Wishing you much consolation.
> Susie Oglesby

MATTOON, ILLINOIS
NOV. 22 – 64

Dear Mrs. John F. Kennedy—

It is with regrets and sorrow that has been in my heart and mind the past year that I write you this letter in memory of Dear Husband and Father of your children. Our heart was also broken and torn assunder over the great tradgedy that happned to him one year ago today.

The reason that I have not written you was because I am poorly educated and was ashamed to write you altho down in my heart my prayers + thoughts were with you and your family.

I am colored and 65 years old and John F. Kennedy is the only man that fought a mighty battle for the freedom of my race of people that they might have their equal rights here in America, which has been lacking since Abraham Lincoln But I consider John F. Kennedy much greater than Lincoln.

Because he was going all the way out for my people. That is why he is sleeping in Arlington National Cemetary to day. The Enemys of my people slew him, because he was fighting and working that we might have our just rights and freedom too live and dwell like all the rest of Americans, and I can truthfully say that the Negro race of people are yet Bowed in sorrow for this Great President who was not afraid to speak his mind + thoughts also to work too carry them out, + he gave his life for us. We bow our heads in shame and sorrow, and wish that we could bring him back to finish the task he started, our world would be such a happy place to live in if we had more men and leaders like him may God rest his weary soul + may he arise to meet the great King whom shall be comming back to earth again to geather his people unto him, and all sorrows and tears shall all be wiped away. Mrs. Kennedy I have been ill eighteen years with a serious heart condition. I am not able to get around much but God has so wonderfully blessed me and spared my life When the Doctors said I would not live till morning

I try to comfort others in writting letters and praying for them and doing good to all both foes + friends and scattering deeds of kindness every day and I only wish that there was some way for me to help lift the terrible burdan and heart ache that I know that you are still carrying

May God bless you +
Your dear little ones is
my prayer for you + may
God ever lead You in his
Paths of rightiosness is my Prayer
And I believe if President
Kennedy could speak to us
He would say
I shall pass through this
World but once

If therefore there be
Any kindness I
can show
or any good thing
I can do, let me do it now
For I shall not
Pass this way again

                        God Bless you,
                        Sincerely
                        Mrs Andrew Burril

Dear Mrs. Kennedy

How are you and your children. I truly hope you are fine. There so much I want to say to you and your family

I am hurt deeply hurt because of our great lost.

But oh how I thank God that he left you here to take care of Carolina and John Jr To lose their father was a great great lost but we thank God that they have you.

I pray that you will continue to be brave as you are lovely

We Negro love you and your children as we loved your husband and pray God that he will keep you well and safe

We also feel as God send your husband as our Moses came after almost a hundred years to open doors for us. doors that had never been opened before I am a witness to these things.

He was water in dry places. He was a shelter in the time of storm also a good servant for our Lord He wanted good for everybody

We know God choose him to do these things for my people The Negro opened his mouth and President Kennedy spoke the words for us.

If it had been in the Negro power President Kennedy would have lived forever. We all loved him but God loved him best, you can truly believe he is in heaven along with your other three children.

Not only President Kennedy but you and your whole family gave us great joy Every day we got the paper to look for you, Carolina or John Jr. so we could see how the first family were getting along.

I realize we have a good man in President Johnson but the Negro of these days will always remember your husband as <u>Our President</u> from the smallest baby to the oldest grandmother

I hope you will find happiness as the year go on

Remember we are praying for you, your children and your inlaws we hope you will pray for us.

So again I say.

I am deeply hurt.

But we still have you. So we have a lot to be thankful for.

Sincerely yours,

Mrs. Marzell Swain

NOVEMBER 23, 1963
PINEHURST, NC

To The J. F. K. family.

Dear beloved one I know you are some what suprized shocked with the Last + Griefed to what What happen to such a great man husband father an Son an Father of our country—but in this Sadist moment of your Life you have my greatest symphy for every one an I know you are suprized to know I am a Negro woman but as I set here an try en put in to words my feeling I hope you Will feel a spot in your heart for me. This marning God spoke to me it said P. J. F. K. did for his Country what God did for his world they Killed our Lord an Father. An now they have Killed our Presentend an Father. We loved him but God loved him best. I feel our Lost is heaven gain how greater man is this to give up his life for his friends. Please dont say he is dead Lest say he is at the waiting room Wating for God to say you have been ruler over a Few things but now you are ruler over many

When his daughter & Son walks through life People will Look on them an say there goes a great girl & boy and a sweet beloved wife but I Feel our Lost is heaven gain.

Please dry your tears dear ones an say I beloved one is heaven gain he isnt dead he is Just sleeping in heaven. I want to meet p. J. F. K. in that great day an help him sing

They Will be peace in the Valley[.] for me I am Praying for the family

Please tell the children to be Sweet an O Bey their mother an Father is waiting to welcome them home

your truly

Katherine Dowd Jackson

A Negro woman with a

Big White heart

P.S. I wont [to] wish Presentend Johnson a happy sucsess in the Pre-densence Field I am Praying evry thing will work out fine Thank You may God Bless an Keep him from harm

---

My Dear Mrs. Kenndey.

Heres hoping you accept my letter. And In Doing so, May God For Ever bless you and Children.

To me, on T.V., Your husband was a God Sent Man. He warm, true and In my heart was and alway will be my Mr. President.

For it was he, As Mr. President strickly opon the way for the Negro.

The trouble to Day, is caused by so-Little Faith in God.

In the next Forty to Forty-Five Year A Negro from Louisiana will be come President of United States of American

Mr Johnson 2 to 1 over Mr. Goldwater.

Watch Mr Robert Kenndey Climb UP.

A Negro

Who beleave In God.

Dear MRS Jacklen

I am Riting you Be Cost I Dont no But I Field like as much as I think
of our Presedent the more I am prest about it I am a colord lady But he
seam clost to me as my own and he was apart of all Armericains I no tho
all is a part of Him Mrs Kenedy to you and your to Children my love and
prayers to all of the Rest family god be loving Child god it and he take
it that when we be gin to think to our self we going along so well and all
at once some thing run acrost our path and HURT US verry Bad so we
must carry on till we are surtly God is with us be Cost if we trust God he
we wont never Be a lone Present Dent was a nice sweet kind man a good
leader and nothing seam to warry Him nothing. But a smile and a Hand
shake now wish you and your faimly all the luck and to Caline and little
Johnie all the love I have I am 74 years you no your Farther was my friend
and Every thing to me to the world love for you

My name is Martha Ross

A mong the tactics used to prevent African Americans, poor whites, and
Native Americans from voting were poll taxes and literacy tests, the latter often selectively imposed on individuals who turned up to cast their ballots.
The 24th Amendment, ratified in 1964, had prohibited use of the poll tax in
federal elections. The Voting Rights Act of 1965 included a provision directing
the Department of Justice to challenge the poll tax's constitutionality. In 1966,
the Supreme Court ruled in Harper v. Virginia State Board of Elections
that state poll taxes violated the U.S. Constitution. At the time of President
Kennedy's death, however, these restrictive practices still remained common in
many parts of the Deep South. Several of the following letters make reference to

these methods of political disenfranchisement as well as broader tensions over civil rights and desegregation.

THANKSGIVING DAY 1963

Our Dear First Lady,

I would not be so presumptuous but a heavy heart compels me to do so.

We are so thankful today for having known and admired our late President. His devotion to God and to his fellow man was an inspiration to us all.

This was my first time to buy a poll tax. I am not proud of my negligence. Our President made me aware of my responsibility as an American and the privilege I so long neglected.

I had planned to thank him by giving him my vote next year. In this day of thanks it is most appropriate to thank you, Mrs. Kennedy for your willingness in every undertaking and for sharing a great American with his country.

May God bless you and your children.

Mrs. American Citizen

FLORENCE, S.C.
JANUARY 17, 1964

Dear Mrs. Kennedy,

You were sweet and moving on television acknowledging the hundreds of thousands of letters you had received since President Kennedy died. It must be astonishing to you to realize that there are people whom you never met, who did not know your husband except at long distance, who wake up now and again at night and weep for their loss. It is almost unaccountable: President Kennedy was too young to be a father figure to people, like me, in their fifties, and too old to be identified as a son. But why, after all, should we try to analyze or explain the emotions that bound so many of us to him. There is comfort in the fact that human personality can hold such strength, that worthiness can so shine forth, diverse people will pledge allegiance to it.

I want to tell you this little story. When your husband was the Democratic candidate for President, I was living in Chapel Hill, North Carolina, and was trying to help in his election. One of my chores was to drive people to the polls to register. Since I worked, Saturday was my house-cleaning, grocery-shopping day and by five o'clock, I was bushed and disheveled. But it was the last day for registration and when I got a call to pick up someone before the books closed, I poked some potatoes in the oven and took off. My destination was a small house on the edge of town in a "Negro section." A neat, elderly woman, crippled with arthritis, sent one of her grandchildren to fetch her husband from the field where he was working. When he got to the car, he said that he wanted to wash up and change his clothes. I tried to be patient as I explained that he needn't do this, that we didn't have much time, and "Look at me—I've been working all day and I don't mind going to the registration place like this."

He was not happy to cut his preparations short, but compromised on some details, and we set off. On the way, he confided that he hadn't voted since the last Roosevelt election. "Mr. Roosevelt put me on the WPA," he said, "and I've been a Democrat ever since, but I haven't felt the spirit move me to vote again until now." Why now, I asked. "We need Mr. Kennedy," he said simply.

We got to the registration place in time and there were, fortunately as it turned out, very few people on hand. He sat down and stared at the printed form which had been handed to him. "You sign your name here," I pointed to a line. He picked up a pencil and carefully made an "X". The registrar and I looked at each other. "Can you write your name?" she asked. He shook his head. "Can you read what you're signing?" she pursued, though we both knew the answer. He sat with his head down. She turned to me. "I'm awfully sorry but you know registrants must be able to read and write." We were both upset—and he was so ashamed.

On the way back I told him Mr. Kennedy would be proud to know that he wanted to help, and we talked about how his grandchildren were doing in school and how important it was for them to stay on and get as much education as they could. He felt rejected then, and undoubtedly felt bereaved last November.

All this about us and nothing about the way your life has been changed, and the lives of your children. You and they so brightened our national scene. Why shouldn't I say to you what I would say to a good friend: that I hope you will marry some day someone who will give you and the children love and companionship. There are many who care about you and want you to be happy again.

Sincerely,

Mary F. Nies

(Mrs. Frederick J. Nies)

··· ❋

ST. PETERSBURG, FLORIDA
NOVEMBER 25, 1963

Dear Mrs. Kennedy,

A few weeks ago while the civil rights battle raged at its highest pitch and the emotions of most of the South were boiling, I made up my mind to write a letter of encouragement to our President. However, like so many other citizens, I am a procrastinator. The letter was never written.

Although I am not naive enough to believe that my letter would have reached a man so burdened with world and national problems, I did have hopes that some far-removed secretary would open it and at least add its contents to the favorable side of the ledger. Ever since President Kennedy's death last Friday I have felt that, by failing to write, I let him down in a specific way. No matter how small a factor my letter would have been, it would have represented an effort to stand with my courageous leader, harassed and beleagured in his fight for justice. Multiply my procrastination by that of thousands in the Southland who must have sympathisized with his efforts, and our neglect takes on the proportions of tragedy—especially now. In a covert way we are guilty of desertion in the face of the enemy.

It's too late to address a letter to John F. Kennedy, President of the United States. I can only join the remainder of our country in openly recognizing and honoring his greatness after its embodiment has passed from hearing. Nonetheless, I take some small comfort in the hope that, someplace, seated in the presence of <u>his</u> Chief, Mr. Kennedy is conscious of these words and that he will forgive me for failing to uphold my responsibilities as a citizen while he bore the crushing burden of the presidency. Further, in his memory, I pledge that no successor of his will ever again go lacking for my "drop in the bucket."

Please accept my deepest sympathy in your grief and my solemn admiration for the noble courage you have exhibited during these terrible days.

Sincerely,

George T. Davis

P.S. A copy of the previously unwritten letter as it had formed in my mind is attached.

St. Petersburg, Florida

November 25, 1963

Dear President Kennedy:

I have little hope that this letter will come to your personal attention. Yet, its carbon copy will serve as a reminder to me that I have made this small effort to let my Chief of Sate know that, even in the section of my country where bigotry, hatred, and resentment are strong, there are some who understand, some who stand with him.

Born in Alabama and raised in Northwest Florida, I have come up a mute observer of the practice of "keeping the negro in his place." I remember most vividly my own negro mammy and her unconditional love for me as though I were her own son. Her love remained steadfast in spite of her own hopeless situation, a houseservant for life, denied by her economic and social status the privilege of remaining at home to rear her own children. I remember the ramshackle negro school in my hometown, its <u>status</u> <u>quo</u> taken for granted because it was a negro school. I remember a white bully's knocking a negro down and kicking him while he was on the ground because the negro had dared to "sass" me. And I remember the election day when the news was grapevined through town that some "niggers" planned to attempt to vote. An almost illiterate white citizen came to the polls with his shotgun and let it be known that its double barrels would be directed at the first "nigger" who tried to enter the place of voting. No man, not even the high sheriff, challenged the right of this self-appointed guardian of white supremacy to deny citizens of the United States their Constitutional rights. Little did I realize then that my indignation toward these injustices was not something to be ashamed of. On the contrary, I had a sense of guilt about it, a feeling that I was disloyal for disapproving the <u>status</u> <u>quo</u>.

Your gallant fight, in the face of outraged, entrenched privilege has been a great inspiration to me. I can see now that what was needed even back then was some one of the privileged who would stand up and rock the boat. I thank God for a President now who has the courage to do just that.

I hear people cast accusations. The more prejudiced they are the wilder the accusations. Some say it's for the negro vote. Some say it's for personal power. Some have even hinted communism. But I know better. I know the burning sense of indignation toward injustice which drives you on. I know that, politically, there is a great deal more to lose than to gain. I know that the real gain cannot be envisioned by crass prejudice and that it can only be measured by the ideals set forth in our Constitution and, before that, by the God of brotherhood. It may be that you will be blocked, but the cause for which you fight can never be thwarted—because it is TRUTH. You have helped crack the shell and a little light has appeared. Men have seen it and, no matter what the political tide brings, things can never be the same.

Thank you for risking your political neck for something which can bring you so much vexation and so little obvious reward. I just wanted you to know that you have one admirer, a son of the deep South, a registered Democrat whose ballot with be marked Kennedy and whose friends will know shy.

> Sincerely,
> George T. Davis

WEDS.
NOV. 27
ROCHESTER, NEW YORK

My dear Mrs Kennedy,

I wish so much that good might come out of this dreadful evil—as a <u>memorial</u> to your wonderful husband. Would the Senate & House pass the civil rights bill at once?

One hundred years ago, Lincoln died that all men might be free
One hundred years later J.F. K died that all men might be equal.

> With deep sympathy
> Mary W. Bentley
> (Just one of the mourning crowd)

JAN 15, 1964
MONROE, LA

My dear Mrs. Kennedy,

After two months, the hurt of losing our great President doesn't seem to ease—and I just ask the Dear Lord each day to comfort you and the President's family so that as days go by things will be easier for all of you.

For me—in my beloved Southland—where there is so much fanatical racial hate—I shall try each day to do everything I can to make the lot easier for the colored people I come in contact with at my work. Their pay is so small and their housing so terrible. My maid cried for days after the assination—she knew from the moment it happened that a bright light was gone from the world—one that we all needed so desperately!

Tell your precious children that the world loved their father—that he was brave and strong and that he loved America so much that he was willing to give his life if things could be made better for all man-kind—and now it is up to each of us to—"Do unto our Neighbors as we would have them do unto us"—Then & only then—Our Blessed Lord will not have died in vain and neither will your dear husband have died in vain.

Thank you so much for your shining example for all women to follow. You too—are a shining light!!

Love & Admiration

Mrs. Linnie Boner

PLEASANT GARDEN, N.C.

Dear Mrs. Kennedy,

Death is hardest on those who loved and are left behind. My heart broke with yours when your beloved husband died, and I grieve with you who shared and created life with him, and with his parents who gave life to him. Yet he lives in the two dear children you both loved, and who remain to comfort you. His life was shared with them and his spirit lives in them. In the memories of all who knew him, either personally or through the intimacy of television, he will also live.

Although I did not always agree with the politics of the Democrats nor with the authoritarian demeanor of the Attorney General, President

Kennedy had the warmth of personality, the earntestness of speech, the compassion of manner that seemed to extend the invitation to peace, love, and friendship to a world needing just such a way to follow.

You too have been an inspiration to all of us. My deep admiration for you is such that my limited vocabulary is unable to describe it. 'Tis said that the happiness of our years as we grow into maturity is measured by the happiness of memories stored in our hearts and minds. You may truly have treasured hours to comfort you as days go by.

I was always proud of President Kennedy. Perhaps the speech which meant most to me was that he made to urge all of us to see that every human being would be assured of dignity and equality of opportunity. This has been a matter of deep concern to many of us, and we were grateful and proud that he had the courage to express the humaneness of it.

I am glad, too, that you decided to let the nation show its high regard for a vigorous, devoted young man who gave his life for his job—the task of leading the people he loved and who love him in return. I pray that the causes for which he worked—human dignity, peace, conquering of prejudice and poverty—may be strengthened through the efforts of all men everywhere.

You are so lovely that I am sure the future will in turn be beautiful in many ways for you and the children.

Sincerely yours,

Catherine B. Parsons (Mrs. J.S.)

ROBERT T. CATES, M.D.
JACKSON, MISSISSIPPI
DECEMBER 2, 1963

Dear Mrs. Kennedy:

As one who loves America second only to God and reveres and honors the high office of president of the United States, I wish to take this means of expressing to you my personal grief and profound sympathy in your time of sorrow. The heart of every decent American is saddened and ashamed at the outrageous act of a warped and twisted mind, which took away your husband, our President. As you know, we in Mississippi have always held opposing political views from those of the late president but in the tradition of Robert E. Lee and Jefferson Davis, we have held these differences as civilized gentlemen and not as depraved killers. We held many more differences and saw our women and children suffer at the hands of the Carpetbaggers and the Negro Legislatures immediately after the Civil War and at that time there were many more sharp-shooters in Mississippi and in the South than there are today. Nevertheless, no attempts were made on the lives of any of these people with whom the South so violently disagreed. We got them out of office by ballots—not bullets. In spite of having our cause misrepresented on so many occasions, we believe that this country should return to constitutional government in the tradition of Washington and Jefferson, and that it must do so by way of the ballot box and no other way.

As a physician who loves life and who has dedicated himself to preserving it, as a husband and a father of two children, I deeply sympathize with you in your loss and in your grief. You and your children should indeed be proud of our late president's courage and of his ultimate achievement for this country. As tragic as his assassination was, it is my conviction that he accomplished in death what no other president has been able to accomplish since the days of World War II; namely, the feat of uniting this country behind our government and our American way of life. I believe America today stands more united and more solidly dedicated to our liberty to the preservation of our republic than it ever has in the memory of most people who are living today. Indeed, it is not the length of a man's life which matters, rather it is how much he has accomplished. I believe this final accomplishment of your husband's will go down in history as one of the greatest. I believe also that it will be a turning point for our nation and will open up an entirely new vision of strength, patriotism, and the willingness on the part of every American to make whatever sacrifices necessary to preserve our country and our way of life.

In closing, I wish to commend you and your children on the absolute dignity, strength and grace with which you bore your tremendous loss. Indeed, God has been with you through these trying days and it is our prayer that He will always give you strength to bear whatever lies ahead.

> With every good wish for you and yours, I remain,
> Yours very truly,
> ROBERT T. CATES, M.D.

.... ✽

PELAHATCHIE, MISS.
DECEMBER 2, 1963

Dear Mrs. Kennedy:

While, I did not approve all the methods used, by President Kennedy, in the handling of Civil Rights, I long ago became convinced of the rightness, of his viewpoint. I find it a source of profound sorrow, that, so many of us, here, in the South, allowed ourselves to become so engrossed, in the racial issue, that, we completely failed, until too late, to note the full import, of President Kennedy's great vision. By being so narrow-minded, we lost forever, the joy of seeing, at least a start, of some of his most important hopes toward ultimate fruition. It took the humble homage, of so many Foreign Nations, to at last, bring us to a full realization, of our loss.

May his great spirit allways rest in peace under the shielding arms of the God he served. May you, in the days ahead, console both yourself and your children, with the fact, that, President John F. Kennedy will live forever, not only in the hearts of his countrymen, but, also, in the hearts and affections, of all the world. May God bless and keep you and your children safely, allways.

> Sincerely,
> W.J.B. Daniel

.... ✽

HYATTSVILLE, MD.
DEC. 6 – 1963

Dear Mrs. Kennedy—

Mere words cannot express my grief over the tragic, senseless loss of your husband and "my president", two weeks ago. I cried as if I, too, had suffered a personal loss—My tears still fall and my life is sad—I am saddened because I am a native of Louisiana, and my grief is caused by the fact that this crime occurred in my beloved south & was caused by a southerner "spawned by the Devil"—The tragic event came on my daughter's birthday. Nov. 22nd, she was five—Too young to understand why her Mother prayed & wept at the news—Her party was somber and she could not understand why Mommy cried when she cut her cake & opened her gifts—Maybe as she grows older, she will look at the scrapbook I'm preparing & understand but I'm glad she was spared the grief—

My southern relatives often wrote and scathingly called your husband "your president"—I'm proud now that I countered their views with the facts—I agreed with your husband that all men deserve equal rights—I replied to them simply—"When Negroes cut themselves, they bleed, too." Many times I also replied that "There but for the grace of God, go You"—They were right only in the fact that they called him "my president"—He was & still is—Maybe, its because, we are of the same generation—I'm 37 & my husband 9 days older than yours—There was a feeling of kin-ship. Will you accept this simple Verse from Shelley that is so very true? Will you also believe me when I say that I will always up hold all that your husband fought for and continue to teach my children what I always did, that God created all men equal—I hope that my deep shame that my be-loved south was the scene of this tradgedy, will one day be replaced by compassion for the blind who will not see—May I say in closing that as long as I live, your courage, beauty, and charm will always keep you "First Lady" in my heart.

My prayer for you and your two lovely children is that the loving

Father who loves us equally and watches over us all will sustain you in your grief and one day, He will again give you the gift of re-newed hope and the ability to see the beauty that He created—Maybe, I, too will be allowed to live my life again, as it was before we lost "my president"—

My deepest sympathy & all respect,
I remain,
Madge E. Asselta

·· ❊

ENTERPRISE, ALABAMA
NOVEMBER 27, 1963

Dear Mrs. Kennedy, Caroline and John-John,

This is the hardest letter I have ever written, and it is with deep regret that I do so.

Although I am a native New Yorker, I am now living in the South, amongst the hate and bigotry that people like you and yours try so hard to alleviate. After a year here I have come to know the bitterness that so called human beings harbor in their hearts. Even so, your wonderful husband and father, did manage to get through to some of the people down here. Those at least, who have the depth to understand that people are people regardless of their color or creed.

What I want to tell you is how I will always treasure the memory of a cold, windy day in Allentown, Pennsylvania, in October 1960. After waiting for more than three hours with my fourteen month old daughter, our future president finally arrived. The ovation he received was of course, no more than he deserved. After making a memorable speech, without showing any annoyance at the baby girl named Stacey, tugging at his trousers throughout his speech, he received the handshakes of the people who could reach out to him.

He held Stacey in his arms right after this picture I've enclosed was taken. He asked me her name and age and the names and ages of my two

Stacey Simrin is the child pictured in the lower left wearing a "Kennedy" banner.

older girls. He took the time to tell me what a beautiful baby she was, and how much she resembled her mother. He wished us luck, autographed Stacey's banner and shook my hand.

This great and good man gave five precious minutes of his time to Stacey and her mother. He proved to all those present that day, what we felt about him from the start.

Little Stacey is now four years old and has always referred to President Kennedy as "my President Kennedy", since that day in Allentown's Center Square. She has in her youthful way, felt deeply the loss all the world should feel. The loss of our great, courageous leader.

Your wonderful dignity and unselfishness during the five days past is a lesson in courage we all will never forget. My admiration of you is beyond expression of words.

If when I visit my family in Washington, D.C. in the future, I would deem it a great honor indeed to be able to meet and speak to you. You are indeed a true first lady.

Please accept my deepest, heartfelt sympathy. Try as I might, I cannot find enough words to express my feelings. I have tried to do it in the only way I know how.

May the light that burns at his grave never cease to burn in the memory of all those who knew his greatness.

Very Sincerely,

Arlene Simrin and family

In the assassination's immediate aftermath, few letter writers expressed much curiosity about who was behind the murder of President Kennedy. The arrest of Lee Harvey Oswald just an hour and a half after the fact no doubt allayed such concerns—at least initially. Only a very thin file of letters reflects any speculation among those who wrote to Mrs. Kennedy about a conspiracy. Most who did spin theories blamed Communists at home or abroad for perpetrating the crime. But many more letter writers described a political climate within American society that they believed encouraged hate. The extremism of those who would fan the divisions in the nation, some letter writers suggested, emboldened those who were mentally unbalanced and might well have contributed to the horrific attack upon the President. Several pointed to a pattern of lawlessness and aggression growing out of racial tensions in American society, some carried out or sanctioned by "respectable" members of the society, as an incubator of violence. There were letter writers who placed responsibility for this state of affairs on the South. But many more saw the difficulties as national. Whatever the source, a great many Americans expressed shame that such a terrible crime could happen in the United States. The most common question posed by letter writers was a simple one: Why? They offered a variety of answers.

A copy of the following letter to the editor of Cornell Daily Sun, *the student newspaper at the University, was sent to Mrs. Kennedy.*

12-2-1963

TO THE EDITOR

CORNELL DAILY SUN:

AS WE THINK ABOUT THE ASSINATION OF PRESIDENT KENNEDY AND OF LEE OSWALD, IT SEEMS THAT AS MEMBERS OF OUR SOCIETY EACH OF US BEARS, IN TWO WAYS, PART OF THE GUILT FOR WHAT HAS HAPPENED.

OSWALD WAS TRAINED BY THE MARINES, IN THE NAME OF NATIONAL SECURITY, TO KILL MEN. HE WASGIVEN A MEDAL FOR DEMONSTRATING HIS ABILITY TO DO SO. HAD HE KILLED AN ENEMY HE WOULD HAVE BEEN GIVEN ANOTHER MEDAL, BUT WHEN HIS PERVERTED MIND DECIDED THAT THE ENEMY WAS THE PRESIDENT, THEN IT BECOMES A DESPICABLE CRIME. WE SHOULD ONLY WONDER THAT THIS SORT OF THING DOESN'T HAPPEN MORE OFTEN.

THE SECOND POINT IS THAT THE GENERAL EMOTIONAL CLIMATE IN THE NATION TODAY, PERHAPS MORE PREVALENT IN THE SOUTH, BUT CERTAINLY NOT EXCLUSIVE TO THAT AREA, ENCOURAGES THIS KIND OF RESPONSE BY IRRATIONAL PEOPLE. STATE GOVERNORS REFUSE TO COMPLY WITH THE FEDERAL LAW; PENNSYLVANIA HOME OWNERS ATTEMPT TO FORCE NEGROES FROM THEIR AREA; POLICEMEN BEAT, JAIL AND FINE YALE STUDENTS WHO ARE HELPING TO GET NEGROES TO VOTE IN A MOCK ELECTION; CORNELL FOREIGN STUDENTS ARE HARRASSED AND BEATEN WHILE TRAVELING IN THE SOUTH, COURT ORDERED SCHOOL INTEGRATIONS ARE MET WITH HYSTERIA BY WHITE PARENTS; AND NATIONAL LEADERS ARE SPAT UPON. ALL OF THESE ACTIONS ARE COMMITTED BY SUPPOSEDLY NORMAL, RATIONAL CITIZENS, YET WE HAVE THE HYPOCRISY TO PROFESS AMAZEMENT AND HORROR THAT SOME OF OUR LESS RATIONAL CITIZENS, INCITED TO

VIOLENCE BY THE ATTITUDES OF THE RATIONAL, ARE CAPABLE OF
SHOOTING A NEGRO LEADER, THEN BOMBING A NEGRO CHURCH, AND
NOW ASSASSINATING THE PRESIDENT

WHEN SUPPOSEDLY RATIONAL PEOPLE RESORT TO DEFIANCE AND
REBELLION TO AVOID ACCEPTING IDEAS AND CUSTOMS THEY DISLIKE,
THEN AN EXTREMIST, IN IN ORDER TO BE EXTREME, MUST RESORT TO
EVEN MORE VIOLENT ACTIONS. DALLAS OFFICIALS HAVE DISCLAIMED
ANY INFLUENCE OVER OSWALD BY THE RECENT DEMONSTRATIONS
IN THAT CITY AGAINST OTHER NATIONAL LEADERS, BECAUSE THE
DEMONSTRATIONS WERE INSPIRED BY THE RADICAL CONFLICT WHILE
OSWALD APPARENTLY COMMITTED HIS ACTIONS AS A COMMUNIST.
BUT EMOTIONAL ATTITUDES AFFECT THE MIND IN ALL AREAS; HATE
ISNT SELECTIVE. WHEN WE AS INDIVIDUALS PERMIT OURSELVES TO
HATE ANOTHER PERSON, THEN WE ARE GUILTY OF CONTRIBUTING TO
THE EMOTIONAL ATTITUDES THAT MAKE POSSIBLE DISASTERS SUCH
AS PRESIDENT KENNEDY'S DEATH.

IN ADDITION TO THE DESTRUCTIVE EFFECT THESE ATTITUDES
HAVE ON THOSE WITH THOSE WITH WHOM WE COME IN CONTACT,
THEY ALSO WILL EVENTUALLY DESTROY US.

DONNA SOOBY, GRAD.

JANUARY 1964
CHICAGO 13, ILLINOIS

Mrs. Jacqueline Kennedy

The sorrow of my wife and Myself is beyond words but I feel I must
try to express our feelings.

We are not the best educated people In the world. I myself have only
a grade school education myself.

However I did learn enough to have <u>Love</u> for my <u>country</u> and <u>respect</u>
for my

<u>President</u>.

This terrible thing that happened I Beleave is due to lack of the proper teaching

On the part of parents, and Government Agencies Alike. and most of all the

American People them selfs.

Sorrow is not enough or just going To church to say a prayer and then leave. Go home and say we did our share.

He the President is gone now and there is nothing more we can do and Forget it ever happened.

We the American people have been Like this for a long time forget the Bad things that happen to us and Only remember the good.

We <u>must</u> never <u>forget</u> what has <u>Happend</u> To him.

He must not have died in vain for If he did all is lost for everyone. And our Country will become the Kitchen where hate is permitted to Rule instead of Justice. And where terrible Plots against one another are allowed to be cooked up.

To take a man's life is the unforgivable. But to let it happen and then in a few days forget about it is even Worse.

You are the first Lady of the land and as his wife if you decide to Take up the fight to <u>wake</u> the people of this country up and teach us your Way of respect and of Honor.

It I think would be the greatest Gift to him, and the people of the Country.

Our greatest need in this country today Is not huge Army's or large stores Of gold.

Not fine homes or tall buildings. We do not need more weapons or Bigger ships.

We the people of this Great country Do need men like him and Ladys like Yourself to lead us and to guide us. We need your energy and honor We must be tought that hate and Greed can only be met with more of the same and neaver end unless we Make it end.

His <u>leadership</u> tought us not to hate and tought us how to forgive but now

He is gone.

But to me and I beleave to all people of the world he will neaver be forgotton.

And must not be for all our sake's.

Respectfully and Sorrowfully Yours,

Mr. & Mrs John J. Floodas

•• ❋

Dear Mrs. Kennedy,

It is Sunday.

Tonight; alone, I contemplate the many events in this grief-filled week-end.

Sad is only a word, and yet, it _is_ the word.

I, as many others, never met this man, and yet I have wept. The love I felt for all that was his way is not easy to speak of now.

A sense of anger has swept me, a sense of uselessness, a sense of sorrow, but always the question, Why? Why?

Greatness was his; no, greatness _is_ his, this man young in body; but mature in mind.

Now I can only question. How can God take from this great nation this man of destiny as he had only begun his crusade of equality and justice for all men.

Is this the Lords way of striking at us for our sins as a nation? Or is this his way of emphasizing the great truths that this man stood for.

And now this man of peace, lies in peace.

In some way may the grief of his beloved wife, devoted family, and unrealizing children be shortened with knowing in his brief life-time he has done more toward greatness than most men could do in many life spans.

The Lord has granted him a place in history, let no man deny this.

With all my heart felt Sympathy.

John L. Touchet

···

NEW YORK
JANUARY 17, 1964

Dear Mrs. J. Kennedy,

The faces of all of the people reflect the deep sorrow the astonishment the ugly ness of the brutal news.

All of a sudden our late and loved President, becomes some one who belonged to us all, and the mind turns in vain to understand <u>why</u> and <u>how</u>

Should it have happened in a hostile country an explanation could be found. But to have lived bravely through the struggle with both the enemy and the cruel sea, to die under the sun of Texas at the hands of an American it is too much for the intelligence to understand and the heart to bare.

And all of a sudden we wonder if we have a soul or if we only breath.

I am a mourner of his death, his picture is on my desk.

I join you and your children in your grievance.

Sincerely

Mrs. F. Storll

···

NOV. 25, 1963
ALLEN PARK, MICHIGAN

Dearest Lady,

I felt compelled to write to you tonight. Although my letter may be only one among thousands—perhaps millions I felt I had to do something personal in this your and the nations hour of bereavement. I am just a homely man—a father of three—a school teacher—a catholic with our home in the vast mid-west. I have spent the day like so many other millions of Americans watching the ceremonies from Washington; so often during the day I longed to reach out to comfort and help you—and this letter is all I can do. I and my family can however; pray—pray for you and your familys comfort and for the soul of our John Kennedy—this we shall do.

I am enclosing a copy of a letter I have sent to Senators Hart and MacNamara and Congressmen Lesinski and Staebler

Sincerely,

Thomas H. Runnals

November 25, 1963

Dear Sir:

On this the eve of the burial of our dear President John F. Kennedy, I feel compelled to write to my senators and representatives in Washington for the first time. I would prayfully ask your support and efforts for any pending legislation on control of firearms. The more restricting the legislation the better. Why do people need firearms today? We have no Indians to fight or savage beasts to protect ourselves from.

Possibly the obvious answer to the question is to protect ones self from another person with a gun—a fine situation for this, the most "civilized" of nations. Would that we could impound every firearm in the nation and make it a federal offense to have one. If the Constitution needs changing in this respect, lets change it! If there is a cause for us all to feel responsible for our nations recent loss it is my firm belief that this is the crux of it all.

If no legislation is pending do all peace loving Americans a favor and introduce something to help protect us and our families from armed men who have <u>absolutely</u> no need for firearms to begin with. Law enforcement agencies should have

strict control of all firearms—hunting rifles should immediately be impounded after seasons—please help us.

> Yours truly,
> Mr. Thomas Runnals
> Allen Park, Michigan

TACOMA, WASH.
NOV. 22 1964̶3̶

Dear Mrs. Kennedy:

Like the whole world, I too have heard of your husbands death,
(our honorable President). I too am grieved over his assassination.
How could any one kill another? It's barbaric, sickening, and just plain
heartless. I do not know the president, nor have I ever met him in person, but I KNOW he is our president. Like myself, I am a catholic, and
like the whole world, I am adding my prayers to pray for his soul, for
you and your children.

I have two children, and I know how you must feel, being left with
them, and not seeing your husband any more. I know I would feel like I
too would want to die.

God Have Mercy On Your Husbands Assassin:

What is now to become of our world, with out our good president?

May I add one more thing Madam President?

Though there were things my husband and I disagreed on, that your
husband thought was right, I could never take my revenge out in the way
that someone did. I respect your late husband tactic, and I'm sure that he
thought the decisions he made were in good faith.

I feel every one in this world, are some what responsible for the death
of J. F. K. We are ignorant people and very hostile, and very prejudice. In
stead of forgetting, Forgiving, and loving thy brother, we fight, and kill
what we think are our aggressors. I have nothing more to say, except, MAY

GOD LIFT THE BURDEN OFF YOUR SHOULDER, AND MAKE
YOUR TEARS A PILLOW FOR YOUR HUSBAND HEAD.

> Grieviously I remain,
> Mrs. Patricia Koop

P.S. If I spelled your name wrong, please forgive me. I am writing this letter with tear stained face and shaking hands.

BALTIMORE, MARYLAND
NOVEMBER 22, 1963

Dear Mrs. Kennedy:

There is no comfort to offer a lonely wife and fatherless children. May
we thank you for everything you have given. It cannot be in vain. No petty
difference or principle divides us on the value of human life. Behold a guilty
and shamed nation. Pray for us someday as we pray for you and yours now.

> Sincerely yours,
> Betty Keat
> (Mrs.)James Keat

BRANDON FLA
NOV 27, 1963

Dear Mrs. Kennedy

I am a Florida Dairy Farmer who has been a lifelong Republican. I
am also Protestant and have been anti-Kennedy since 1960.

However I feel a desperate urge to exend my deepest sympathy to
your children and to you. As an American I'm deeply ashamed at the
manner in which the President met his end. As an American I'm also
very proud of the great courage you displayed during the entire ordeal.

I have talked with many of my friends who have like back grounds and I can assure you their feelings are 100% the same as mine. We are so sorry.

May God be with you in the days and years ahead.

Sincerely Yours

Russell E Weir

SHIRLEY, MASSACHUSETTS

NOV 22, 1963

Dear Mrs Kennedy

This is the saddess day of my life would mean noting now. But let me tell about my self my name is Roger Lewis Burke Jr. I work as a dish washer at the Park Street Dinner 1 Park street Ayer Mass. I work for a fine man Lester W. Berry I am 20 years old.

I am taken a high school cource at home and later on I hope to go to the Harvard College in Harvard Mass. I am doing this for jack so so I can someing for my world. I never knew my father my mother gave me away when I was 1 day old to my grand parene Her name was OBrein. My mothr kill her self in Prov. Rhode island 1956

Jack was murdered because the earth we live on is full of vices in crimes grat corruption and evil

Jack is in paradice and he is looking down on us waiting to see iff all of us had the faith he through we had or we are nothing but amimals

Dear, Dear lady you are not alone ever man woman child on this univerce are with you all the way and look to God and remember

He shall dweal in the House of the lorde forever and ever

some time in the future write to me and tell me that you are all right.

Roger L. Burke Jr

POLITICS, SOCIETY, AND PRESIDENT, 1963   131

Dear John-John,

I'm sure you've been watching what has been going on these past few days before your third birthday. I saw you on T.V. the other night as you saluted that big black box as it went down the street on its cart. Because of that, I decided to write to you to tell you why you won't see Daddy anymore.

I'm sure it's very hard to understand—even I can't believe what's been going on in front of my very eyes.

What is going on in Dallas now is very ugly. Cancer is a disease that spreads through the blood and lives off the very important parts of the body until those parts are so changed that the person dies. It's sort of the same way in Dallas. A cancer is down there, too. It's not something that you can cure, because it's hate. I'm sure you've felt hate once or twice, but after awhile it goes away and you feel better again. The hate in Dallas does not go away. It's called "blind" hate— and that's the cancer that's flowing through the bloodstream of America, poisoning it's organs and ruining it's functions until it dies of this disease.

Your Daddy wanted to see this cancer. He went right to it's center to see if he couldn't make the people be nicer or at least make them find out why they are hating. But I guess they didn't have time to find out. Maybe it's because Daddy didn't even have time to tell them, because before he could, the cancer began wrapping it's tentacles around him, the tentacles of hate, and discrimination, and bigotry, until the squeezing got too hard, and Daddy slumped in his seat.

You won't ever see Daddy, but many part of him will still be there. You won't find them by looking with your eyes for them, but, when you get older, you will know how to look another way. It is called looking with your soul. You will find many, many parts of your Daddy which he left for you to take by looking that way. Many people cannot look that way, and so cannot find the things to take and to love. I hope you find them and take them.

You will understand much more when you grow up, John-John, but right now be nice to Mommy, because she does understand.

These are the things I wanted to write to you about, John-John, when I saw you salute that big black box.

[NO SIGNATURE, PLACE]

321376
CENSORED
COOK COUNTY JAIL

To the First Lady

Mrs John F. Kennedy

To you I take this opportunity to express my feeling towards the greatest American ever lived, most of all your pride for one such as Mr Kennedy, should never be forgotten. We've lost a great fighter, and most of all one which have the knowings of his fellow man. As we read in our Heavenly bible, earth has no sorrow that heaven can not heal. Mrs Kennedy I must say you're a great soldier, one of strong feel, you are to be honor for your outstanding manner in which you've held yourself during the tragic hour; you are quite a Lady. May heaven ring many a blessing around your path. Prayer is the key which shall unlock the door, and faith shall embrace it's wealth. Your little lady Caroline, all I can say she's a darling, the world loves her madly, you have children which you can be very proud of. If you notice my tribute to your husband isn't of fancy work, yet it is written from the very depths of my heart. I am no one special, just an inmate of Illinois.

Yours Truely
Ricco Barnes 321376
Cook County Jail E-1
Chicago ill

A Tribute to Mr J.F.K.
You the congress of these United States and offices of leadership
In the late President John F. Kennedy, yo've lost a very magnificant man
into the arms of earth.

His pride and dignity, none could ever surpass, his warmth was one
never to be forgotten.
Leaving behind millions to weep, both national and international
Also leaving our traitors to joy at our sudden lost, declaring their
friction in
our land of great Freedom
Oh how it must hurt, Mr and Mrs America's hearts, to know that their is
slaughtering going on in these United States of America
Killing each other as tho they were Texas cattle in a pen
Oh God !! bless our nation, and protect our leaders along their path
From men whom dignity have no cause.
Spray your loving grace around Jackie and her tots . . .

ERIE, PENNSYLVANIA
MARCH 16, 1964

Dear Mrs. Kennedy,

I feel that I too must join in the chorus of sympathy that has been showered upon you in regards to your husband.

You must not grieve great lady, because that is the job of our country. The world may be divided into vast empires, cities split in half, countries fighting against against themselves, and neighbors killing each other, but all men of every race, color, and creed came to give you something. It may not be a gift that can be seen, but one that you can feel in your heart.

The reason that all men morn the death of your husband is because we feel that we have played a part in his murder. We may not of been there, or even known that you were there but we are all responsible because we did not fight with him in everything he stood for, we just sat back and watched. We have wasted a great man.

We may not realize that fault now, but in time, history will write its own story on blood filled pages.

We all feel the loss of him dear lady and I as one am sorry.

Eileen Harayda

My dear Mrs. Kennedy:

Enclosed is a copy of a letter my family recently received from my brother, Lieutenant Thomas A. Michalski.

Perhaps you may never even get to read it, but nevertheless feel that it strongly evidences the thoughts of our family in regard to the assassination of your beloved husband, and our beloved President.

I personally, could not have expressed more sorrow and grief had it been my own dad. My only consolation is that he has met his Heavenly Father and Eternal Reward, and is now being repaid for his goodness during his short, all too short, visit on earth.

My love and fond remembrance in prayers is with you, John-John, and angelic Caroline. Mr. Kennedy will be remembered in my Sunday offerings at Mass forever.

Sincerest condolences,

Lorraine J. Michalski

Dear Mom and Dad, Sis, too

My God! What is happening to America? Just what are things coming to? The barbaric murder of our President is unthinkable, unbelievable! I was at the Rhine-Marn in Germany and a desk clerk said that Kennedy was shot. I didn't believe it! I thought someone took a pot-shot at him as they did to Truman.

When the report was confirmed I headed for the Officer's Club. Ladies were coming out in tears and I knew it was fatal. The silence was terrifying—a moment of prayer! I was completely shaken then, quite frankly, and unashamedly in tears as I left the club. It was inconceivable that he was dead. I couldn't believe it! I went into Frankfurt. The Germans were completely excited, "ERSCHORZEN! DREI SHUSSEE KENNEDY TOT!"—Three shots, Kennedy dead. I wandered around aimlessly, stunned, and I'm just now coming out of it. I still can't grasp it, I just can't. Poor America, whither the Union! My heart was torn with grief.

The human grief expressed over a young man's death, admiration for the magnificent heroism of the First Lady was evident. The Requiem In Aeternam just cut into the chambers of your heart.

Government moves forward, but the breakdown of the Constitutional System, bombings, riots, assassination, has me very concerned for America, despite my optimism for her.

Kennedy was in a very real sense, though a pragmatic politician, urging a very real war for the American dreams and their fulfillment. In this he was alone, vertically alone, trying to balance out the forces of extremes. The north, south, whites, blacks, hate, violence, stubbornness, and he did it in an artful, open and cheerfully confident manner. There was a sense of shame we must all bear in this. It is not to say he was always correct. We are all fallible in that we supported him in all things, or did not have certain misgivings on certain policies. This, too, is altogether human and American, but God! not by the bullet through the head. Not by riots and bombings.

They killed a man, but also an image. For above all Kennedy was the American success story—this boy who made, President, the young, vigorous, leader who embodied to the world, and this was so evident in Europe, the best features of America.

It seems as if this has been more than an offer; one's life snuofed out. It is the fact that all the evil in America, life, race, class, sectionalism, radicalism of the left and right joined in destroying the good. Evil has triumphed. God grant it is a momentary thing, a passing triumph. That is what cut my soul so deeply.

Kennedy the person was the son of the immigrant, the catholic, the former officer, the well classically educated north-easterner, the humoristic, the young man, the man under whose call I received my commission during Berlin, the man under whom I served during Cuba and that lonely hill in Germany, where we were literally one step from hell on earth.

I hope the ghost of this man haunts us for many years to come, even as that brooding statue of Lincoln does. I hope this crime chills the very marrow of the American body politic. Not since Lincoln's death has there been a crime of its kind, so heinous, so completely revolting to human sensibility, so tragically devoid of reason.

The handling of Oswald was stupid; it can't help but give rise to every sort of doubt and suspicion. I wonder if we shall ever know the truth.

The great idealism of the Kennedy Administration must not die—the odd effect of marriage of the idealistic, young, educated, positively prognostic was a tremendous achievement. I hope Johnson keeps the core, if not the accents of the old administration.

The whole nightmare is something out of Macbeth or a Greek tragedy. Kennedy has stood before the judgment of God. America and her place in history are yet to be finally judged.

Requiem in aeternam, JOHN FITZGERAL[D] KENNEDY. God Bless America. It seems from here we need it.

As ever with love, your son,

Tom

MR. & MRS. TEMPEST C. ZINN
STEWARTSVILLE, NEW JERSEY

Number one.

1. Your husband helped the colored people for which I was glad and thankful for.

___

2. Your husband was a good man.

___

3. and trash came along and shot him.

___

4. We are white people.

___

5. We are formerly from Pennsylvania. our homes section is over on the other side of Harrisburg, P.A.

That is the way we go to it from here. We moved to New Jersey.
Spring of the year 1953. from and Signed

Mrs. Tempest C. Zinn
Stewartsville, New Jersey

___

extra
Abraham Lincoln helped the colored people to. and trash came along and shot him to.

___

I know what mean people are like. we live right beside mean people. it is our Boss and his wife. and they don't associate with colored people either. What is color? No black ever rubbed off on me. God is no respector of persons. and colored people are welcome in my house any day they wish to come.

EVANSVILLE, IND.
NOV. 22ND, 1964

Dear Mrs. Kennedy:

How are you and the children? I hope my Letter will find all of you fine, under the circumstances. I want you to know that you aren't alone in your SORROW. I, Like Millions, of others are sharing it with you. The death of your dearly beloved husband and our dearly Loved President, is a Loss, that can never be restored. I Am a Negro and I just Loved President Kennedy. He did so much for my Race. There's Lots of Negroes that wouldn't have the Nice jobs they have if it hadn't been for your husband. He Proved that he beLieved that all men should have equal opportunities Regardless of their Creed, or Race, by the good he did for all manKind. His being Assassinated is something I'll NEVER get over as Long as I Live. Our American officials are always trying to tell other countries how to Live and yet in this modern age and day, an American citizen Kills the President of the United States. It is the terriblest thing that could have ever happened to us. I for one hang my head in Shame. After our President was Killed in cold blood, a citizen was able just walk up and kill his ASSASSIN, as far as I;m concerned no-one was on the jobs those heart-breaking days Last November. Everytime you Look around you see Resturants, or Filling Stations being built, those aren't the things they should be building. What we need is hospitaLs. The world is just full of Oswalds Running A-Round. When doctors and officials discover that a person is suffering from insanity, they keep them in confinement awhile and then turn them A-Loose on Society. The problem is that we just don't have the hospitaLs to hold all the sick people. In my opinion this is one of our worst problems, because it's not saFe to put your children outside to play, a woman can't walk down the street alone, and a man isn't safe either. So until something is dont about the mentally ill People all our Lives shall be in Constant danger. I had always wished for our first family of Our Country to be younger people. I got my wish. Mrs. Kennedy you and the President were just wonderful, and your two Little Children are as sweet as they can be. I pray that God will give you the strength and courage that it takes to Raise children trying to be both mother and father to them. May God bless and Keep you.

Sincerely Yours
Mrs. Wilma Brooks

EXMORE VA
NOV 24, 1963

Dear Mrs. Kennedy,

I can't express my sympathy in words nor speech but in my heart you have it all. I have been hurt from the Assassination of our Dearest and best President of the United States but he has gone to rest on a pillow of Gods Abreast, and he will never be forgotten for the good things he has done for us all especially the colored race in which I am.

Grace your self you have something too be proud of a hero man that wasn't afraid of the world the only man that could conquer the whole Soviet Union and keep peace within the whole United States. God has his reward more than we could ever give. I know the hurt you have because it not only hurt you but the whole United Nations, but you just keep smiling. We loved him, but God loved him best and gave him a home of eternal rest. May God ever bless you and your children and may they grow up in years to come to see that they had a father to be proud of. I am poor with a family of 10 but I knew if President Kennedy had of live I would of been rich not in money but in spirit. He was a good man an well loved by the Colored race we all are hurt. If it wasn't for "hate" he would have been living today. "Hate" is the meaning of all of the world brutaly slaying of nothing but innocent people that love to [do] good. So let us that love love one another no matter what color the skin is we still love you because from hate we have lost a Dear member that seemed as if one of our family. I didn't have a TV but I watched a neighbors' and seen the funeral. I gave you credit for being a loving, sweet, and brave woman of a wonderful lover. We were crying just as hard as you were. My children asked me, why did it had to be President Kennedy. I am a woman poor and I know what hurt is. So in my last message I said it to you in his own words, "Lets not forget the pride and the Joy of our Fellow man" and in my words is God shall wipe away all tears. From a colored family that Love your husband from the depth of our heart. He shall never be forgotten.

Mrs. Doris E. Baines

Even as they recounted their sense of loss, Americans of all ages discussed the lasting impact that President Kennedy had on their views of American politics and on their own goals as citizens. In so doing, some expressed their views on what they considered to be the pettiness of politics, lauding Kennedy as a leader who rose above the ordinary. Immigrants who wrote condolence letters seemed especially drawn to Kennedy for exemplifying the freshness, hope, and possibility they sought and felt they had found in coming to America. Refugees who fled from fascism wrote of their belief in the President's capacities for world leadership. Other immigrants expressed appreciation of JFK's public support for an end to restrictive immigration quotas based on national origin. In July 1963 Kennedy had proposed to Congress legislation that would end the national origins quota system that had been in place since 1924.

Just three and a half months before his death, Kennedy published an essay on immigration reform in the New York Times Magazine. *The August article drew upon a 1958 pamphlet he had written as a Senator, which was published by the Anti-Defamation League of B'nai Brith. Noting Emma Lazarus's famous words on the Statue of Liberty—"Give me your tired, your poor, your huddled masses, yearning to breathe free"—he observed: "Under present law it is suggested that there should be added: 'as long as they come from Northern Europe, are not too tired or too poor or slightly ill, never stole a loaf of bread, never joined any questionable organization, and can document their activi-*

ties for the past two years.'" Current immigration law, he argued, erred not in restricting immigration—he did not favor open immigration—but because "many of its restrictions are based on false or unjust premises," Kennedy insisted that American immigration policy "should be generous; it should be fair; it should be flexible. With such a policy we could turn to the world with clean hands and a clear conscience." After Kennedy's death, his earlier writing on immigration appeared as a book titled A Nation of Immigrants *(1964). President Johnson steered to passage in 1965 an immigration reform bill that incorporated Kennedy's proposal.*

DECEMBER 4, 1963
NEW YORK, N.Y.

Dear Mrs. Kennedy,

The public ceremonies are over and life goes on inexorably. Only the dull pain remains in my heart and will not go away.

Fully aware that you have received hundreds of thousands of written expressions of condolence, many of which from the most prominent people of the world, I am confident that in your graciousness you will bear with me for giving in to the impulse of adding my modest words.

I came to America after enduring persecution under Mussolini's regime, witnessing the horrors of the war and having my mother snatched from me by Hitler's gas chambers when I was only slightly older than your Caroline. When I arrived here I was without a country and my only wealth consisted of hope and dreams of a better future. Five years later at least one of my dreams came tru. I became a citizen of the United States and for the first time in my life I had a country which I could proudly call my own, a country where I had been able to make a good life for myself.

I volunteered my services enthusiastically during the 1960 campaign to elect Mr. Kennedy to the Presidency, regretting only that the demands of my job did not permit me to do more. I rejoiced with millions of others upon his victory and for the following three years his integrity, his brilliant leadership, his vigorous determination to fulfill the promises he had made to the people and his immense courage to the bitter end inspired me to become a better American, while your impeccable behavior as a wife, mother and charming hostess and as an ambassador of good will through the world inspired me to become a better woman.

It is as a woman that I am now tempted to beg your forgiveness for having helped, even though in the smallest of measures, to place your husband in the position which led him to his cruel death. As an American I ask only that you accept my everlasting gratitude for having allowed me, the Nation and indeed the whole world to share this great man with you for a little while.

I pray that the supreme sacrifice of John Fitzgerald Kennedy will remain an indelible memory in the mind of every man, woman and child everywhere on earth, serving as a bright guiding light to a better world.

I can only hope that you and all the Kennedy family may find some measure of comfort in the sure knowledge that all humanity weeps with you in this dark hour.

God bless you and your precious children.

> Respectfully yours,
> Judy Sheldon

Dear Jackie,

I wrote this a month ago but hesitate to send it. I felt you might not like to be reminded but now I force myself to send it this time as it is 6 months now since Mr. Kennedy died & it is a remembrance of him. So excuse me if I have trespassed my boundary but I want you to know that I care too & Remember.

> New York, N.Y.
> April 4, 1964

Dear Jacqueline,

Day after day I have the intention to write to you. This is "at the back of my mind" constantly. Today is Saturday and I thought of reviewing all the magazines and newspapers piled up in my apartment. You, see, eversince the President's passing away, I have saved every newspaper and magazine that contains the slightest reference to him or his family. Today as I pore through these papers I felt that I must take up my pen <u>now</u> and write to you.

I bought a Kennedy scrapbook from the five and ten and I will paste the photographs and important articles about the President and about you and your children and about the Attorney General.

I just want you to know that we think you and your family are precious to us.

We are Chinese. We feel that President Kennedy is the only President who takes an interest in immigrants. I have read excerpts from "A Nation of Immigrants."

John F. Kennedy is truly a great young man afire with love and service to his country and to the world—to humanity. He can truly, in his own words, "turn to the world with clean hands and a clear conscience."

Consider the words Emma Lazarus wrote on the pedestal for the Statue of Liberty, "Give me your tired, your poor, your huddled masses, yearning to breathe free." We are the tired, we are the poor, we are the huddled masses yearning to breathe free, and John F. Kennedy had tried in vain to help us. And now even he has left us—the flotsam of society. But we, the flotsam of society, take much courage from him. I want you to know that because of your dearly beloved husband, we are <u>better</u> people today. Take me, for one example. Eversince the President left us, I have undertaken tasks which I have not done before, to better conditions at home, in the office and in my Catholic Church parish. I said to myself, "if John F. Kennedy did so much with such tireless energy the least <u>I</u> can do is to try my best in everything I do." And that is the first step toward building a better America.

Now I am beginning to think of myself not as a Chinese but as an American. I realize that this <u>is my</u> country. We are fellow immigrants. We are gathered here in the great American Experiment and it is up to us to make it a success.

I am preparing myself now. I shall study further about American history & civics so that when they call me to take up my American citizenship papers I can truly say in my heart that like John F. Kennedy I love my country.

My friends say I am silly to write to you, that you will never read my letter, that your long line of secretaries will read this & just throw it away and send back a mimeographed form letter with a stamp of your signature. They are of little faith. Faith is tremendous, for faith can move mountains and if Mohammed will not go to the mountains, the mountains will come to Mohammed! God bless you.

Love,

Mary Quan

P.S. My sister Dorothy has her first daughter named "Jacqueline" & we call her Jackie, after you. My sister lives in . . . Gardena, California.

P.S. To take one day out of our lives—& to tell you about it.—Take the one month anniversary of your husband's death. During Sunday Mass I prayed for you & your family & for the President. I remember the Chinese family seated one pew in front. The little kid brother was opening his teen-age sister's prayer book & many religious cards (Estampitas) fell from the prayer book. Then I noticed that the "In Memoriam" card with President Kennedy's picture also fell off from her prayer book. I also have a prayer card with the President's picture on it and a prayer for the repose of his soul.

Also I want you to know that many women did not feel like dressing up in Easter finery this year because they said that since the President's passing, they lost interest in foolish spending for vanity. There are many other tales I want to tell you for you to take a little consolation & comfort that we also care for you & feel for you in your hours of

Be sure to phone us in advance if you intend to visit China town! . . .

We like to treat you to a Chinese dinner! My Husband is Frank Quan. To the Chinese Rathskeller Restaurant where my husband Frank is working as head waiter. You will see that Chinatown is a happy busy town with Chinese & Italians living in Peace together.

THURSDAY NOV. 28, 1963

Dear Mrs. Kennedy:

Twenty-six years of:—escaping from Hitler—growing up in wartime China fleeing from Communism—watching my father's futile struggle against cancer—seeing my roommate killed in an automobile accident—all these I deemed adequate preparation for some of life's bitter moments.

Yet <u>NEVER</u>, until last Friday, have I felt such a desperate sense of loss and loneliness.

Yes—"A piece of each of us died at that moment."

Yes—"I dare say we shall never see his like again."

& Yes also—"When those who have never met us weep when we die we are truly loved."

President Kennedy was, and always will be—loved—by people the world over.

So from one—a privileged foreigner—privileged to be given the opportunity to live in the midst of so much greatness.

In humility and gratitude to a beautiful man who irrevocably restored the world's faith in the human race.

My Sincerest Sympathy—

> Respectfully,
> Gabriele Gidion.

12/2/63
DES PLAINES, ILL.

Dear Mrs. Kennedy:

It is now ten days since the tragedy which overcame you and all of us. It is fitting that I should write to you and tell you of our—my wife's and my own—deep sorrow for you and your children. You have lost not only your husband, but we all have lost so very much. We are both immigrants to this great country, coming to Boston from Israel in 1947. We were both born in Germany and had to begin twice anew in our lives. We have 3 girls 5-8-10. I have been unable until now to really associate myself within myself to this new and different country, which is only natural I suppose. The reason is quite simple. I had never <u>felt</u> an overriding patriotic interest in our congressmen's and senator's actions, or our presidents for that matter. They all seemed more interested in potatos for Maine Oranges for Florida, cattle for Texas etc. Most of them stood for the status quo, for trading and give and take (I vote for your bill if you vote for mine). That is probably a very superficial impression, but that is all you can read if you are on the sidelines. All this changed when Mr. Kennedy was elected and started his high office. His dedication, his purpose, his vision, his leadership, his realisation that there is more to the President than just "to carry on", was a source of inspiration to us. He had the courage to not only see beyond today, but to try to lead the way. His unfailing optimism that eventually people will follow him to a new frontier was electrifying. After having had leaders like Ben Gurion who had, I believe, many qualities similar to your husband, or Winston Churchill (I served in a jewish battalion in the british army for 6 years) who had no peer at all, I had felt let down here in all these years. This terrible tragedy suddenly made me to identify myself very closely with America the beautiful. We shed tears for days, and I visited a synagogue for the first time in many years to pray and think. Your husband will be missed by thousands of people for years to come because he gave them something to think about, some path to follow, somewhere to go, even though they may not have been ready or willing at the time.—Last not least I cannot but say what a source of inspiration you must have been to him. Having been blessed with a wife who is everything a man can ask, I can so readily see what you must have meant to him. Your devotion to your husbands public image, your sense of history during these tragic days, your inner strength has the admiration of all of us who admired you both. I think you should, in memory to him, and as an example to us all, remain eventually in the public eye in whatever form suits you best. Let me thank you for having given us inspiration and hope that from time to time there will be great men in our midst.

Sincerely
Peter Brenner

John F. Kennedy was among the wealthiest Presidents in American history. He came of age during the 1930s when the United States was reeling from a decade-long economic downturn of catastrophic proportions with massive unemployment. He admitted to a journalist in 1960 that he had "no first-hand knowledge of the depression." "My family," he explained, "had one of the great fortunes of the world and it was worth more than ever then . . . I really did not learn about the depression until I read about it at Harvard."

Despite his personal affluence, elite education, and privileged social circle, Kennedy earned the affection of many Americans of modest and very little means who believed he cared about their economic prospects in American society. Early in his presidency, Kennedy advanced increases in the minimum wage, an expansion of Social Security benefits, and extension of unemployment benefits. His wider objectives—federal aid to education, support for expanded mental health facilities, an antipoverty program and health insurance for the elderly—never reached fruition in his lifetime, instead becoming landmark achievements of Lyndon Johnson's presidency. But Kennedy's insistence on the national scope of these problems, the responsibility of leadership from the federal government, and the moral obligations to redress the situations of those Ameri-

cans most in need resonated with countless citizens. In the flood of condolence letters came many messages from men and women of few or modest resources as well as those with physical challenges who nonetheless believed that he was working for their best interests.

JAN. 19, 1964
BREAUX BRIDGE,
LOUISIANA

Dear Mrs Kennedy,

I'm writing a letter to tell you how sorry when we hear the die of Mr Kennedy.

Because it hard to hear someone is kill. It was very hard to hear kill.

Because we suffer just like it was our on brother. It very hard for you and your childrens, too.

Mrs. Kennedy I use to hear every time he was talk on the television, I use to lession every time. I wish I could have none how to write fass. So I could have write every word he talk about. But I can not write fass. Because I want only in the 4th Grade.

He had send a letter to mine daughter Betty Lou Albert She had the letter Saturday after he was kill. She did'nt answer the letter. She was so upset of the kill. She ask me to write to you Mrs. Kennedy. I hope you don't mine.

Mrs. Kennedy we are country peoples. But we was so fine of our President. And all of you and childrens, too.

He want'nt every childrens to go to school any how they are poor.

We are poor but I want mine childrens to go to school.

Mrs. Kennedy all pray for is soul every day.

Well Mrs. Kennedy I will close mine letter. But not the love we had for your'll.

May God blesses you and your childrems.

And lost of luck, too.

From
Mrs Clytus Albert & fly

Lawton Oklahoma
December 5, 1963

Mrs John F Kennedy & Family

Dear Mrs Kennedy we are
a family of seven poor never
have any of the real advantages
of life always hasten pinch pennies
to pay our bills, but we try
to live a respectful life.

We loved the President as a father
as the father of our Country.
No one can never take his place.
We also know he couldn't have been
so wonderful if it hadnt been for
you. being such a wonderful wife.
With sincere sympathy I will
close with love.
Mr and Mrs Ralph Cohee and
Family

Wishing you a very pleasant
Christmas and New Year.

Mrs Jackolyn Kenedy & children

Dearest ones

Just a line from me to you. Wondering how you and children are. Fine
I hope. Mrs Kenedy God Bless you I want you to know that you have our
deepest Symphty You gave up the greatest man on earth and you were so
Brave and taken Everything so wonderful Even to your Darling little children they were so brave I never can see how you could be so sweet about
the whole thing I must say you taken it better than my family did I have
been sick since the day it happened Dear you'll never know how much
and how often you and children are thot about. President Kenedy was
loved by the world He was the Greatest Leader we ever had or ever will
I will never for get when we heard what happened This was a very sad
home and we miss seeing him and hearing him as yet very badly. I never
felt so sick in all of my life as then and have been very ill since then in fact
still in bed for the last 3 weeks. I must say again he was the greatest man
on earth. I will never for get the night he was elected My husband & I sit
up all night with him when the Battle was won I went to bed with all ease
was so happy But just as sad as I were ever happy when the tragic came
We never had the privlige of meeting him in person but he did seem so
dear & near. It could have never hurt any worse if it would have been one
of our immediatly family I've wanted to write you ever since it happened
but I just couldn't bring myself to it. Wish you didn't live so far away I'd
love have you and children come to visit us. My daughter is a Retarded
child she is 21 Never been to school My Husband is in very poor health
he haven't very much left in Stomach he has had 5 operations will be 7
years in Oct since he has been able to go to work the dr. say he will never
be able to work again he draws 40.00 a week worksman compensation
which is a very low salary for 3 I work when I can in summer in canning

factory am over weight and feel petty lousy most of the time we had a
nice 90 acre fruit farm until sickness hit my husband then we practically
lost 60 acres and has had to mortage home quiet heavely seems as if it
take every penny for taxes, interest, insurance. school taxes are so high
still there is no school for my Daughter which makes me feel very blue
sometimes My Daughter can not be left a lone She is a constant care we
love her very dearly only wish I could do more for her. She is very sweet. I
have to go to work when I can and sit and worry about her all day knowing I should be with her and my husband. Still its impossible I think
from year to year things will be better but nothing in sight Well Dear
you'll never know how boaring it can be and I think I have my troubles
but I look around there is always some one else I guess is worse. At least
I have my husband and Daughter with me if it is a poor way to go some
times. Don't no how long I'll have my husband Dr's doesn't give us much
courage I told my husband I get so disgusted. Wish there could be a
change as Day in Day out is some pretty long Lonesome days for Carolyn
to have to be away from me. It is terrible and hurts to the bone that little
Carolyn and John caint have their father and such a wonderful father
May God Bless them and you always.

Respectfully yours

Mrs. Riley Woodwrick

Some day we hope to see President Kenedy Brother Pres. he looks so
much like your husband. when younger. Pres Johnson has certainly done
wonderful also was so nice to have some one who could take over.
P.S. Oh! Say I for got to tell you my husband & I use to get such a Kick
out of President Kenedy when the News Reporters used to suround him
with questions all he had to do is just open his mouth the Answers just
flowed out. he never had to study for a minute or any one else that had
questions for him. I still say he was smartest man on earth and is greatest
missed not only by a few but the whole world. Even to Carolyn Sue when
he came on T.V. she would run out after us to hear him she has lots of his
pictures she cuts them out of magazine she always Referred to him as the
News. She still cries and talks about him Well Dear it is a long time to

wait to mail a letter and card but as soon as I could bring myself to it. May God be with you and children always.

If you should have a little time please let me hear from you some times

Earlene

JANUARY 16, 1964
MCCLELLANDTOWN, PA.

To Caroline and Little John,

    May you always remember History will be filled with the goodness your daddy did. But the poverty stricken, poor people who survive can tell you better.

We all loved him.

Joyce Wise

DEC. 25. 1963
SALT LAKE CITY, UTAH
MRS JOHN F KENNEDY
WASHINGTON DC.

Dear Mrs Kennedy

    You have never mett me, and probbely never will.

    And as you are such a bussy women, I hope you will pardon my intrusion upon your time.

    But as I was very fond of your husband (PRES Kenedy) and so I would like at least send my heart felt symithy to you and your children.

    I know that there is no word or deed that can take away the sorrow you have in your heart.

    I myself am just a nobody from nowhere,

    But I do want you to know that we on the lower level,s are having our hearts bleed for you and the Nation.

    When Pres Kenedy was here in Utah last and spoke in the L.D.S.

Tabernackle I was in the balconey and look right down on his notes I was so close to him and certainly enjoyed his talk.

I am an Engineer, Power House of the L.D.S. Hospital,

And I know that we all feel the same toward you and your family as I do.

Very Sincerly I remain your unseen friend.
Ernan H. Smith

P.S. You are so bussy. I do not expect A repply. EHS

....❈

SPRINGFIELD, OHIO
DEC. 3, 1963

Dear Mrs. Kennedy

You don't no me. But I am one of you and your husband's great admirers.

Mrs. Kennedy next to God the president was the best friend the poor class of negroes & the poor whites had. Because he opened up doors for us. Especialy though that are [not] to lazy to work. I suppose you wonder why I am writting this letter to you. Because I want you to no how my family & many others appreciated your wonderful husban.

When President Kennedy took his office. We lived in a shack. But I was proud of it because I couldnt do any better. I worked hard to try & keep it clean. My husban didnt have a job. He just work when could get it & had unemployment.

Then he got a better job. Last winter it was so cold in the house there was only one room in the 7 room house that was warm when we went up stairs to go to bed was just like going out doors in the cold. I said lord if I have to live like this another I dont want to live. I started to praying ask god for a better home. And to give my husban & I plenty of work & I our health & strenght to keep our work & a house after we get it. And with gods help & Mr. President we are in a better home a very nice home which I am so proud of.

I just thank god & Mr. Kennedy for opening doors for us.

My husband & I are not lazy. We have 10 children. One girl married one in the U S Navy she is

Miss Ruby J. Givens R.N.

U S N Hospital Staff

New Port, Rhode Island

I am so proud of her is in the Medical corp she is working in Physial Therapy she say's there is nothing so great a feeling as giving a person the use of an arm, leg, hand that has lost the use of it she says it is wonderful.

We are not looking for a big Xmas. We are going try & get enough money to gather so we can have Ruby home for Xmas. Our new home & Ruby if there is more for us in store god touch some ones heart I have 5 boy & 5 girls a wonderful family.

I pray that every precious memories will all way be with you. And make you smile instead of cry for joy knowing you had a jewel for a husband.

Love You
Ruby Givens

....🌸

CINCINNATI, OHIO
FEB 10-64

Dear Mrs Kennedy

Just a few lines as a Friend & Sister of God. I would have written before now but I have ben so filled up I coulden. I will say this if it hadn been for this wonderful man President Kennedy I don't know what I would have done.

in 1963 I wrote the president a letter for viewing my case of help. I am the mother of 15 children I have 6 children at home yet I lost there father in 1959 Social Security diden wont to help with the 6 children that was Elgeable at the time an I diden get it until I wrote to the President he wrote this head office an I got some help the next week after they got a letter from there. These people let me know it. I thank God for this.

I pray every Day an night for you & children as a Sister in God I love you an all an I will say you will always have my prayers. because I love you. Dearly. & hold up your head pray as you go along in life As I know you are a Christian.

God will everymore look upon you as his lite of this world.

Although I an colored but I love you

I will close for this time

> Keeping you & children in my prayers.
> Sign Mrs Bessie Williams

....⁂

ELEVA, WISCONSIN
[JANUARY 16, 1964, POSTMARK]

Dear Mrs. Kennedy and Children

I know this is belated, but after seeing you on television Tues. I couldn't help but take time out and write a few words of Condolence: in hopes you will get this letter.

Knowing you are very busy I put off writing until now, and also time heals all wounds because when this terrible thing happened, it was hard for even us to try and think clearly.

You and your Husband seemed a part of all people including "just us plain farmers," and when my husband heard of the terrible tragedy he remarked "now we have lost our only hope we had," in regards to many problems—including farm problems as it seems it is getting harder to try and make ends meet and now we have lost our Good leader besides.

That fatal nite we went to Mass, in Memory of your late Husband, which made us feel we were helping in our small way by Prayer for him., and so "God" would give you strength through this ordeal, and continue to give you courage.

There were many times I wanted to write to the President—when I would read in the paper of his back troubles,—as I also have had the same experience but, being just plain country folks—I didn't know if it was the proper thing to do, but now I wished I had.

In closing I want to say—you have Our Deepest Sympathy and May God and Mother Mary Help you.

Love from
The Duane Becker Family
Mrs Duane Becker (writing)

MRS. CHA. BINGLE
ROCHESTER IND.

Mrs Kennady,

Words are so, poor of a way to express one's feelings and usually my words never mean much to anyone. It's as if I were a moron or a child and so, if you never read my letter it won't matter much at least I've voiced my opinion if only on paper.

I lost my father in July and while he had no money he was rich with wisdom & I leaned on him. He was a christian man & always made things right with words of the Bible. I know you have lost much more than I, but not really for he was such a good man. With the world growing wicked all the time at least he safe from what is yet to come. He was no ordinery man. While we consider America a civilized Nation we do have animals in human form here too. I know with his wisdom he knew this and went out to Battle a brave man. I know he died for me and my family.

The reason I really wrote is to tell you how I felt toward it all from beginning to end. I had 3 reasons why I resented him to begin with. One his age. I felt he was to young to take the responsibility. He proved we needed a young mind to cope with todays needs in the world. No 2 was his religion. I being raised a Methodest, knew little of Catholics except what I was told. However, as I've grew older I tried to increase my knowledge by examining every thing I didn't know of. Because of Being Catholic, I thought the president would bring it into the government & from all I had heard I didn't agree with it but, anyway I went to Catholic church and explored the people & etc. While it was like every other church I went to (I didn't agree with somethings) it was more like God and the people lived by the religion more than any other church. I found myself liking it. The things folks said were untrue because they never cared enough to learn of what it really was. They preferred to critize the whole church for a few of its members.

No 3. I recented his position and money. All my life, I've never had the things I needed to live. I've worked so, hard and yet all I have is my children 6 of the nicest. They are my desires in life. They deserve bet-ter than me. All I can do is not enough for them to live normally We struggle to keep fuel in winter and when Christmas comes the relatives don't have enough to buy anything for mine each for I have to many. So, our small town gives us used toys mine receive I've never had the money to buy the Groc. I need let alone toys for Christmas I've dreamed often I was in places & I got to take what I wanted in Groc. Clothes and toys and I knew it was a dream. My husband earns $119.00 ever 2 wks & the County say's they can't help anyone with a job. Yet! There are people that for generations back live from the county and are young enough to work who really have more than us. Their medical expense are paid & they receive checks for Groc & Normal things. So, due to my position I recented him but, I found he had values money didn't buy him. He was as human as God himself. No one can take his place, no one has his knowledge, His level headness, his compassion for human race. No words can say what we've lost as a nation and I truly fear for our nation now. The animal that took him from us don't know what he's done to us all. While he done a lot and left us great values. He could have done so, much more.

I have 6 children, 3 boys, 17, 14, and 12, a girl 9, Randy (boy) age 7 who has broken both arms 2 times in past year. 1 time 2 broken at once. Penny age 6 while the small one's knew something terrible had happen my oldest ones felt it with reality of a relative. I had to go get my 17 yr old from school that day to take to eye Doctor and he came out of school crying big as he is (6'2"and said "Mom, I'm determined now to work my self up to being a Secret Agent." I told him he'd never have the chance to guard a man so, good as Mr Kennedy. This boy is going to be a State Police. It has already been planned. He has put himself thru last 3 grades of school by working summer months The president's death has given him added Vitelity to better himself & the urge to go further with his education, no matter the price.

I truly think, you'll always be a first Lady in my heart. You carried your cross so, brave under the public eye and I know your grief must have been unbareable. Yet! The worst is now to face life alone but, you do have part of him. The children they are the symbol of your love for one another and He lives on thru them. I pray God willing give you strength and courage to bare your responsibility as well as grief. No one knows what you've lost only you.

Forgive the paper as it's hard for me to come by also. but, paper can't tell the values of a persons mind

All I can say! is the world lost a Brave & Noble leader and the next few years will show how much we really lost. There will never to be another like him.

Sincerely Yours

The Charles Bingle

Family

P.S. There is so, much more I could say of him but, I'm sure you already know more than I being his wife you knew many Values in him.
P.S. I know of Some who would laugh at me for writing you & tell me how foolish I was but, I don't care.

Dear Mrs. Kennedy:

I don't know whether you will ever see this letter but I feel that I must write it. Only in this way am I able to assuage my grief.

I am one of the "little" people—one who could know the President only through means of the newspapers, radio and television—but his death assumes the same proportion to me as if a member of my own family had suddenly and violently died. First, the unbelief—the prayer that this is a mistake; then the prayer for a miracle—"Oh God, let him live! Let Jackie take him home!—Then, "Oh Father in Heaven, Thy Will be done: when the news finally comes that the President is dead.

I watched you running to the hospital when they carried him in and thought—"Oh Lord, she's just like any other wife." When you appeared at the airport in Washington, stunned and dazed, I prayed for God to give you strength to live for your children.

Since Friday all work has stopped in our household. We get up, go through the motions daily living, read all the newspapers and listen to all news on radio and television.

We meet friends and everywhere the question is the same. Why, Why, Why? The answer is simply this—Almight God makes no mistakes. John F. Kennedy, one of the truly great men of our time, had to lay down his life to open men's eyes and to soften men's hearts to the ideals that he cherished.

He was criticized and rebuffed many times for his ideas of freedom and dignity for all people all over the world; for his ideas that all men need a decent living; that old people need adequate medical care; that all children are entitled to a good education; that all men need equal opportunity for fulfillment of their lives. Because he was wealthy, and had had a sheltered life, he could have led a life of ease, unaware of the needs of lesser people. Instead he chose to carry the weight of the nation and the world on his shoulders.

We, the "little" people, caught some of his spirit—because he was young and vibrant, we were revived in youth and vigor; because he was confident, we became confident; because he was reverent, we became more aware of the presence of God; and because he was dedicated to his principles of justice and mercy, we re-dedicated ourselves in our daily lives in our dealings with others. We were privileged to have lived at the same time that he did. You were more privileged to share his personal life—to share his hopes, his dreams, his triumphs, his sorrows and his disappointments.

God expects goodness from all of us—He demands greatness from a few. John Fitzgerald Kennedy was one of these few. Because he lived with dedication to the service of mankind, we who are left to mourn his passing and to reflect on the meaning his life had for all of us, will rise from our sorrow to carry on in our own "little" way and to try to be a little bit better because of him.

> May God bless you and your children.
> Respectfully yours,
> Ruby K. Nichols

SUNDAY NIGHT
WAUSEON, OHIO
1 _13 /64

Dear Mrs. Kennedy,

My Name is Minerva Chapa. Iam a Mexican girl. I am fourteen years of age. Weare from McAllen, Texas. We are eight in the family four are in school we are all behind in school. Because we have missed a lot of school as you might know my mother &father are labors. THISyear we stayed here in Ohio.

I am writing you to tell you how sorry Iam for what has happen.Ever since he became President of The United States in 1960 Iwas wishingI could meet him and tell him how mush I liked him and was behind him all the way in every thing he did.We are Catholics to but that is not the reason why we liked him so much. It was because of all the things he tried to do, for example THE CIVIL RIGHTS, TEX CUT, & THE STRONG SPEECHES HE SAID AND THE STRONG STANDS HE TOOK, and how he tried his best to do his job and he did the most wonderful job Ithink many mightnot agree with me but this is the way I feel. The world lost a wonderful and great leader.

Isaw every thing on television and Icould not believe it I still do not believe he is dead. I still keep saying President Kennedy instead of President Jonson. Iwant you to know how much I admire you took it i know how you must of felt you knew he was dead. I just do not know how any one could do such a thing such a terrible thing to such a wonderful person like he was. I like you to know I will never never forget him never .Some peaple did not like him because he was rich we our seves are poor but we liked him and admired him for his courage.Please answer my letter. Ifeel so owfle and sad when I think about it. Say HELLO to Carolinne, John JR. and the rest of the family from me to them.

(P.S. We have his famous speeches, we also have three MEXICAN records that describe him so wonderful.

> PLEASE ANSWER? WITH LOVE
> ALWAYS,
> MINERVA CHAPA

SHILOH, TENN.
NOV. 25TH

Dear Mrs. Kennedy & children,

Excuse me for writting a letter but I don't have a sympathy card and I live 18 miles from town and being 71years old and we have no car and I don't get to Clarksville very often. But I want to tell you that I loved President Kennedy dearly and his death has hurt me so deeply that I can't eat or sleep. When he would appear on my T.V. screen his presence and that wonderful smile would brighten up my home so much. How I will miss him no one but me will ever know. I tried not to ever miss him when he was to be on T.V. I have a heart condition and I didn't know if I would make it through his funeral but I ask God for strength and he gave it to me. When he was shot I ask God to not let him die that we needed him so much but he took him any way So he needed him too. With all his wealth he was concerned with us poor people. And why would any one dislike him I will never know. I was borned and raised in Tennessee and there were lots of negroes around at that time but they never bothered the white people and we never bothered them and I was raised to that and I did feel hurt over the Civil Rights not that it would affect me but I have 7 children also grand children and great grand children that it would affect by having to go to school with them. My children are all married. I saw his grave today on T.V. I'm saving all his pictures to put in a scrap book. I get them out of the news papers. My religious belief is "Seventh Day Adventist" And many people didn't like him because of his faith but that never kept me from loving him and I was hoping to live so I could vote for him again next year but if I'm here I'll vote for President Johnson even if he does come from Texas. I can never like that State again. Cheer up Mrs. Kennedy for you have 2 lovely children to give you love and pleasure. And little John looks just like his Daddy. I'm so glad that you was with him as it had to happen. May God give you strength to carry on is my prayer.

Love you
Mrs. Helen Oldham

ATLANTA, GA
JAN. 14, 1964

Dearest Jackie,

I just had to let you know how much my family and Mother regretted the terrible incident of your husband. He died a brave man and was loved by most everyone.

The most important thing I want to tell you is how my mother just seemed to worship your husband. She is nearly 80 years old, can't write her name hardly, no education at all, but has learned a lot from television.

She would listen to his debates with Nixon before he became president and admired your husband. She for the first time in her life was interested in voting but couldn't since she can't read very good or write only her name. It is the first man she even noticed as President, even though we're a poor family.

When you husband was assassinated, it nearly killed her too. She was sick for weeks and at that time she left her television on and listened for anything she could hear about him.

President Kennedy was in the hearts of all the aged and poor even though he was wealthy himself and I know he wanted freedom and peace for all. He reached the hearts of many that no one else could reach but our Lord God.

We are from the South but there were many, many people here that loved him for what he did and what he was.

May you and your children be rewarded in many ways in the future for your having such a wonderful husband and father.

May the unity be between you and your children as you had while he was living. That is something else I admired him for—unity.

You were a wonderful, first lady and was also admired by everyone. You were so brave of an example to your husband at the time of his assassination and death. You are still loved by all and your children are loved also.

We will never forget you, your husband and your children.

May God be with you forever in being a father and mother to your children.

Our Love
Mrs. John G. Cook
And Mother—Mrs. Henry Wood

Mrs. Kennedy,

Just an Indian who was deeply moved an shocked awakening as to things we take for granted from day to day. He carried our burden of this Reservation on his shoulders. I cannot find words to express my family an I feelings, except that we've come to love him very much these past three years as our <u>Great White Father</u>. He was a man helping us here. I know because he was that kind of a man. We've never before known such consideration. My sympathy an prayer's go out to you an yours.

I remain a <u>Freind</u>

Mrs. Pearl A Charley

....❋

Dear Mrs. Kennedy,

I awake at dawn of this day with the grim realization that our president is dead. My words mingle with tears flowing through my pen as I write.

I am a Crow Indian woman who married into the assiniboine tribe and I write without a moments hesitation that the heart of every member of this tribe has been torn out by the roots and placed upon the ground by fate, as mine is. Yet, time shall pick up each heart and shall place it back in our breasts. Gently soothing the torn and broken parts with the wonderful healing balm that our Creator has placed in the hands of time. Ah! but the scars shall remain as long as we live. The sadness and despair of Our First Lady reaches out across the miles to bring tears to the eyes of the strongest heart.

The voice of Our Great Leader, John F. Kennedy seems to fill the air telling us to <u>stand</u> brave and true for the freedom of this great nation.

I pray that the Great Comforter, the Holy Spirit will comfort the hearts of you and your beloved children now and guide you through the rest of your lives through, the <u>Son</u> of our Creator, <u>Jesus</u>.

With the mourning song of the Indian nation still on my lips I write to let you know we care. With my heart on the ground, I join you in sorrow.

Mrs. Alma Snell

DEC. 6–63

My Dear Mrs. Kennedy:

After watching everything on television and not being a very well educated person, I wasn't going to write you, for words I would use would be so small to comfort you The loss of your husband, the children's father, my president whom I so admired and looked to for help. But today the sixth of December just got me, when the President was to visit us here in southeastern Kentucky.

Let me quote John G. Deitrich (U.P.I.) will describe us more fully, "Kentucky and especially its eastern mountains which have lived so itemately with tragedy bowed heads with rest of the nation in grief over the assassination of Pres. John F. Kennedy.

The fact that this commonswealth never gave its political favors to the dead Pres. in no way altered the stunned sorrow of its people. If possible, Kentucky's mourning was deepened by the knowledge that Pres. Kennedy in his last days had come to a realization that something needed to be done for the mountain people quickly. Arrangements were completed for him to visit the Kentucky mountains on Dec. 6 to see the situation here with his own eyes. Now eastern Kentuckians hardened by years of faded promises, cannot but wonder what will happen to these programs under a new administration facing a Congress reluctant to spend the money involved."

You see, my dear Jackie, we begin again hoping, I had written President about federal tax cut for low income brackets, but now I've lost hope in it. My dear Jackie, you were so brave all during everything, being the great Lady you are, it was expected of you. I will never forget little John saluting, the memory of him will live with me always.

The eternal flame burning for our pres. will surely show others the path which he was helping us to make. It is my prayer that you and children will have a lovely Christmas, may God bless you all.

When a loved one is gone, we can't have them back, we know that, but there's one thing we will always have, and that is our memory of them blessed memories, how they can comfort us. A poet once knew these memories and put them in a poem, I'll pass on to you, may be you too, will find comfort in it as I have.

Long be my heart with such memories filled
"As a vase in which roses were once distilled
You may break, you may shatter if you will
But the scent of the roses will hang around it still."
May God bless you allways,

> Most sincerely,
> frances Ralston
> Middleboro, KY

BELLROSE—L.I.

My Dear Mrs. KeNNedy,

My GrandMother and I Are Not of the weilhed peopLe. As A matter oF Fact we have LiTTLe or No schooLinG. BuT After This proFond shock, iT has shook the rich and the poor. We oF the poor Just DIDN'T Know how to Express our sTunned feeLing to Express to the GrEAT LAdy you Are.

PLease ForGive me For beinG Forward. IF This couLD Be jusT A LiTTLE COMFord To you My GrandMother keeps a CANdLE to the BLessed MoThee For our BeLoved JAck KeNNedy The 35th Presed-

eNT OF the UNiTed STATES. We would iF we couLD SAY Is There
ANy ThinG we CAN do. ALL we couLD SAY I AM <u>Sorry</u>

RespecTFulLy yours
Mrs PaTi
VIVIAN MANFRE.

Dear Mrs. Kennedy:

I want to write you at this time to let you know that I for one lost a dear
friend when John F. Kennedy was so quickly and unfairly taken from us.

My name is Arthur John Gambardelli, I am forty-six, have been blind since
I was a baby, and, am a graduate from Perkins School in Watertown, Mass.

In 1953 I had a nervous break-down and not having any finances and
not wanting to be a burden to my partially blind wife, I volenteered to
go to a State Mental Hospital. I am not really bad, I have a slight heart
condition and of course because if it I am affraid to be alone. I could have
been taken care of had I had the finance to pay for doctors. So, I sat here
and dreamed of the things I would like to do but couldn't because of circumstances My life looked pretty dark.

In 1960 I began to see light because I had heard of a friend with personality and convictions of things he wanted to do for his country. Of course
that man was our President J. F. Kennedy. I got a shut-in ballot and voted
for him beca use he was my kind of man. As time went on and he gave his
program for the elderly, the mentally-ill and the blind I found that he and his
whole family were becoming my friends. I was glad to read of our president's
achievements. I was also proud of you and your two little children. President
Kennedy his two children and You have been my friends since 1960. I read
everything I can about the whole family. I would gladly have given my do
nothing life for his any-time. The kind of life I am living there is plenty of.
The things President Kennedy and the whole Family stood for there is

not much of. I was deeply shocked to hear the news of President Kennedy's death As far as I am concerned I lost a dear friend. Also in my heart John F. Kennedy will always be President Kennedy and You will always be the First Lady and of course I will always Pray for the President the two little children and You. Because I feel that the Kennedy family and what they stood for can never be replaced.

I would like to know if you will please do me a favor. I know that I cannot see but I would appreciate having a copy of the picture like those given at the Mass. I will always pray for President Kennedy his two children and You as Long as I live. God Bless You All.

From a friend who Loves You All.
Arthur M. Gambardelli

BROOKLYN, NY
NOVEMBER 24 1963

To the Kennedy family

This is with deep sorrow and sincerest feeling that I wish to extend my condolences to you

You do not know me but I have always liked and kept in close touch with the happenings of the world

I am handicapped and live at the Jewish Chronic Disease Hospital here in Brooklyn New York City I have been an admirer of your husband son and father ever since he was senator

I know that he shall always be rembered by me because he was a fine person a humanitarian who believed in human individual right and to me he was another roosevelt

Let me end this note with this one thought may his soul rest and be remembered in our hearts forever as he died for our country

Sincerly
Natalie Siegel

The West Virginia primary marked a turning point in John F. Kennedy's 1960 campaign for the presidency. Although then reliably Democratic, the state's population was overwhelmingly white and Protestant. Many believed Kennedy could not overcome local suspicion of his Catholicism and background, especially in a state where unemployment and poverty in Appalachia, as well as hardship among coal mining families, joined "politics and hunger."

Some West Virginians freely admitted that they opposed Kennedy solely on the basis of his religious faith. But as he campaigned in West Virginia, and confronted the religious issue head-on in a televised program that featured Franklin D. Roosevelt Jr. posing pointed questions on the matter, Kennedy began to make headway. He devoted much of his time in the televised appearance to reassuring voters that he would honor the separation of Church and State. Noting that when a President took the oath of office he "is swearing to support the separation of church and state," Kennedy observed, "he puts one hand on the Bible and raises the other hand to God as he takes the oath. And if he breaks the oath, he is not only committing a crime against the Constitution, for which the Congress can impeach him—and should impeach him—but he is committing a sin against God." In the end, Kennedy won the primary and forced Hubert Humphrey from the campaign.

PRINCETON, W VA.
MARCH 1964

Dear Mrs. Kennedy, Caroline, and John John

I have been trying to write my letter to you every day. But could not bear to do so. I thought perhaps a little time might help. But it does not only gets worse. My grief for your husband is as painful as when it all happened.

A book would not hold all I thought and felt for our beloved President and his family.

I was left alone to raise three boys and a daughter. I had a struggle raising them. But after Mr. Kennedy was elected President of U.S.A. all my doubts of future faded. All I could think of was a bright and hopeful

JFK shaking the hand of a coal miner while campaigning in West Virginia, 1960

future. Mr. Kennedy gave me strength and courage to face all the tomorrows and years to come. Now all that is gone forever.

I am a democrat all my people are to. I worked for the democrat party during the 1960 election. also the W. Va primary all public places, super markets, any where I happen to be. They did not say anything about Mr. Kennedys religion, that I didn't tell them off. They shut up too. That was the only thing I heard was religion. Now they are sorry. They now bear in the sorrow.

When Mr. Kennedy was campaining in W. Va my children and I waited for the arrival of the Caroline at the Mercer County air port near Bluefield, W. Va. to see him. We waited and waited. At last here it came. We were so excited. It was in the early spring. We shook his hand. He was smiling. We told him our name was "Gore". He said "Seamed to be a lot of people around here by that name" We told him we were all going to vote for him. He smiled and said "For us to get all our Aunts, Uncles, & cousins to vote for him too". We said we would. You were standing in the back ground smiling. Than away you went toward Bluefield, W Va. to start the campaigning in the coal fields.

Ever time I am going that way I look and say to my self not to long ago Jack Kennedy rode by here and looked at these mountains. I fought so vigorously for Mr. Kennedy in the Primary election. Ever one told me I should get a good job. I didn't do it for those reasons I saw a wonderful man for our future, and was tired of our state being poverty stricken with the other party in office.

He gave me courage & strength after he won. I went to Richmond Va. to seek employment. Got a leg injury and came back to Princeton W Va. in Oct. 1963. I think of Mr. Kennedy, of you Mrs. Kennedy and your two beautiful children everday I cry & grieve all the time.

Since he is gone, all my hopes and dreams for the brighter days I wanted so much to see are gone forever no shining smiling faces, Little childrens laughter gone. All the hope I had left for my family and people in West Virginia. My children all live in different states to secure employment.

I looked forward so much hoping they would change the name of West Virginia to Kennedy. All hope of that is gone too now. . . . I shall always believe Oswald had help in doing what he did. I hope & pray it will all come out in the open. . .

I am planning on visiting Mr. Kennedys grave this summer. As the public says a little of us died when he died. But for me not a little, but all of me died. I suppose it was because I was left alone to raise four children. I had no income at all. My parents died. No one to go to in my sorrow. I felt like I had some one to turn to. I could see his smiling face on T.V. in the newspapers. He was there with you somehow. He made us here in W Va. feel like he was sent from heaven to take the place of Jeses. Sometimes I think he was. You know they crusified Jesus so they had to crusify Jack Kennedy.

I do not watch the T.V. news reports any more. . . .

Good luck to you Mrs. Kennedy, Caroline, and John John.

With all My Love

Mrs. Lennie Gore Housley

P.S. Mrs. Kennedy
If it had been possible

I would have given
My life to spare
Mr. Kennedys life

This is Why I BeIieve John Kennedy Went to Heaven

I was one of the few West Virginians who did not vote for John
Fitzgerald Kennedy. My reason was pIain and simpIe. He was a Cat-
thoIic. I was afraid he wouId us e his reIigion in office. When I saw that
he did not I was immediateIy sorry that I did not get on the Band-Wagon
with the rest of my state and support the greatest president we ever had.

I know many have asked and maybe you have too, "If he was such a
great and important man, why did God Iet him die?" Many people are
wrong about death. God has no part in death. Since the Garden of Eden
the deviI has been the father of death. God is the father of good onIy. In
the OId Testament we find that God is abIe to proIong our days if we
Iive by certain standards or that we shorten our days by riotous Iiving.
But I can think of two reasons why God would want John K ennedy
with Him. One is because He is a jeaIous God, and had he stayed here,
pe opIe wouId have, and especiaIy the coIored race, worshipped him
who was doing so much for and suffering so much for them, and for us
poor fOIks. Another reason is, had he stayed here his biIIs would had
lesser chance of getting through. Yes, it's ironic but had he Iived his biIIs
would have died, but he died and his biIIs wiII Iive. Yes, the government
wiII fight poverty such as we have here in West Virginia, and it wiII
fight intoIerence for any human being regardIess of race or coIor. That is
Gods way, That was John Kennedy's way. So great was his Iove of God
and so great was his Iove for his neighbor though thousands of miIes of
water or Iand separated him from them. Who, but him couId bring the
worId together at his death? Who but him couId cause a worId to truIy
mourn his passing to To weep openIy, unashamed, Iike only very cIose

relatives do for a dear loved one. Only his own family was forced to weep in private.

It was the loveliest fall any of us remember I'm sure. We had some of the most beautiful sunsets. Artist couldn't paint them half so well. The President's face was radiantly happy. Some said it was because his wife Jacqueline was with him publicly for the first time since her illness. That was part of it of course, but it was something deeper. It was an inter-glow, reflected on his face. The sun was as bright and as warm as summer. The Pres ident's great love for people kept him going back and back again up and down the fence doing the thing he loved most, meeting and shaking hands with people. Then with top down bes ide his wife he began his ride down the Main Street of Dallas. For ~~the~~ President John Kennedy this had been a perfect day. Then instanly he was with God.

If he had laid and suffered, he who had suffered so much already, I don't think his family, our nation or the world could have stood it. There was only one thought to comfort us. Our fallen hero had felt no pain.

The lovely fall ended as suddenly as his lovely soul. The sun was darkened with sadness, and a determined breeze began to blow, and the angels wept cold and bitter tears. That is why I believe John Fitzgerald Kennedy went to heaven.

Suddenly everthing that had been perfect was im-perfect. The weather was bitter, rain did not cease, the sun refused to shine. The band and Jacqueline stumbled over frozen clods, and Taps was sounded off-key by a quivering lip, and the world mourned his passing.

I hope this will be a comfort to you, his family. It has helped me to write to you the things I feel about John Kennedy. Even though my old typewriter is beat to pieces and I know it is not perfectly written in any respect.

Merlene Snider

HUBARD NURSING HOME
CHARLESTON, W. VA.
DECEMBER 14, 1963

Dear Mrs Jacqualine Kenedy

After such a long delay I must extend to you and your family my very
Deepest sympathy. Wanted to send telegram but not enough money. I
have lost many loved ones. but never any thing that has affected me like
the loss of my Dear Present. he was the finest Present we have ever had
or ever will have. there is no one can ever fill his place. unless it would
be a Kenedy. they are the must outsting I have ever known. and you Mrs
Kenedy are the bravest person I have ever heard of. no one but you could
have stood up as you did through such tragedy. Please kiss little John
John and little Careline for me and tell them this is a Grandma that loved
their Dear Father as much as they did if that is feasible. I dont think I
have ever missed a word he has spoken over the radio, even when he was
in Ireland, during and after the Funeral.

now I will tell a little of myself. I am ninetenty one very nerveus and
little or no edecetien . at present in nursing home very disatified. I like
to have my own little place and will be soon as I can find something the
OPA gives very litle. now I want to tell you something very unusal. A
friend of mine had a memorial service at the largest Department store
here they bought a flat of White lilies of twenty five. then they brought
them to me. I gave them a drink every morning and now that the blo-
soms have faded the lily and foliage is growing and looks as if its going to
bloom again. Every one says they have never known any thing like it. its
so precious to me. now please don't think think me silly excuse mistakes
and pencil as I have no pen

and again please accept my sincery sympathy. and have as nice xmas
as is possible after such a terrible tragedy

very truly

Mrs Emma Donnally

.... ✳

GORMONIA, W. VA.
MARCH 17, 1964

Dear Mrs. Kennedy!—

I am just an old Mountain woman that has lived on a farm all her life. I became interested in politics at the tender age of 5 1/2 years old listening to my father a Republican and my maternal grandfather a Democrat debating the pros and cons of Major McKinley and Mr. Bryan a free silver D[emocrat]. Of course I was very young to absorb all they said. I remembered enough to follow me all my life and keep me interested in a presidential election.

When Maj McKinley was assinated I grieved for him the same as a very dear relative.

When Col. Roosevelt took over and made such a good president my "grief" was over. Out of this all Major McKinley election my father gave me a fine puppie. He paid all of .50 for him. He was, my father always said, one half "Mastiff," one quarter bull dog and the rest "dad burned shepherd." He was a huge dog taking after the "Mastiff" side not fearing man or beast. He was a loveable dog to me a play mate and protector. He was about sixteen years old when he died.

He also gave me a beautiful bay mare for my very own to ride as I pleased. I followed each election with great interest. When women could vote my husband and I went to the polls and voted to the best of our ability.

We were very happy over your husbands election for the presidency and followed his career with great interest and well wishes.

We were very glad when he decided to run for the second term as I truly thought he was the best man regardless of party.

When he left for his speaking engagement with you and party I followed him with great interest. The nite before he was to go to Dallas I dreamed I was with the party and the plane was bombed. There were several explosions. I tried to find the president and they told me they had taken him to the Hospital. That after noon when I turned my ~~Radio~~ TV on they said Mr. Kennedy had been shot. I knew that my dream had come true. Instead of the plane being bombed it was the gun shots I heard.

I have wondered since if I had been a wiser woman could there have been any thing that I could have done to stop him from going to Dallas. I said "No" for that morning I had seen him going out for a "breakfast" speech. I knew he was a man dedicated with a purpose in life. Nothing I or any one else could have stopped him. This was the way that he was to go. His life was a blessing and his going away a benediction to all of us.

I send you and your children my love and symphathy. I hope you will all have a good life.

I never told my son until several days later about this dream. His reply was "Mom if you ever tell this to any one they will think you "crazy". I have these dreams on several occassions which do come true. I dont claim to portray coming event or to be crazy as my doctor could verify this that I am a very sane person. I told my son I was willing to take a "Lie dector test" to prove that I wasn't lying. I am shy by nature.

I hope my dear you will fine some fine man who will make a good husband and father to your children. You are to young to live alone.

I lived with one man for 46 yrs. he died and I find life is very lonely. No one has time for an old lady.

May God bless you and yours and may you and children have a long and happy life. I am,

> Yours sincerely,
> Dora W. Wildesen

John F. Kennedy was the first American President born in the twentieth century. At forty-three years of age, he was also the youngest man ever elected to lead the country. (At forty-two, Theodore Roosevelt was younger than Kennedy when he assumed the office after the 1901 assassination of William McKinley.) Those facts shaped perceptions of JFK and his administration from the moment of his inauguration. Eight inches of fresh snow had fallen early in the morning of January 20; the temperature hovered near 22 degrees when Kennedy was sworn in. "Rejoicing in his youth," one newspaper observed, Kennedy stood without a top coat in his morning suit to take the oath of office. He made a striking contrast to his predecessor—the seventy-year-old Dwight D. Eisenhower. As Kennedy delivered his stirring inaugural address, telling the nation that the "torch has been passed to a new generation," each breath seemed to etch his promises in the freezing air. "The Inaugural will be recalled and quoted," it was then predicted, "as long as there are Americans to heed his summons."

"The new generation," Kennedy invoked was, of course, the World War II generation. And yet his call to service, his emphasis on change, his high ideals and faith in the future excited many younger Americans. Men and women in their late twenties and thirties, college students, teenagers, and even still younger children of the postwar era claimed John F. Kennedy as "their" President. And when he died, they wondered how to integrate the harsh fact of his brutal assassination with their once exhilarating expectations. As one young man put it, the night of the assassination, "It is the legacy of great men, dead before youth is fully spent, to disturb the foundations of our thoughts. For me Kennedy did just that—he showed me in no uncertain terms that if the 'good' is at all attainable it can only be through a total commitment of self. To understand the Kennedy experience," he hazarded, "Americans are going to have to go beyond the examination of events and happenings, and go to a land of ideas and dwell there for awhile. Kennedy's personality was so strong that he tempered the environment of a nation very quickly. . . . I inately comprehended a political force speaking to me in a language I could understand. I now feel lonely without it."

John F. Kennedy delivering his inaugural address, January 20, 1961

BROOKLYN, NY

Dear Mrs. Kennedy,

I imagine each person who has written you has expressed his sympathy in his own way, and I must express mine in my own way, which is to tell you how much your husband – my president – meant to me. It is a selfish way, perhaps, to write a condolence letter, especially in the light of the question of your loss compared with mine. But perhaps I can speak for many.

Many of my generation did not care for politics, because we felt that we could not, through politics, make anything any better. Those ideals which we held sacred were separate from politics and politicians. The frustration at not being able to do anything about these ideals in a direct manner became too much for most of us – we chose instead to write poems, study theatre, and argue about foreign movies. Was there trouble in our community? Poor housing? Unfair taxation? What could we do?

Our parents felt differently – they had lived through many great social changes and had seen with their own eyes the affect of the people on politics. But our generation, I am convinced, had lost touch, and was retreating more and more from the field of public service. It was with an innocent heart and an uncaring hand that I placed my very first vote for John F. Kennedy. Presidents were people who at best did not make things worse, but they were <u>certainly</u> not idealistic, or even if they were, they were not nearly as free to act out those ideals as was one of the college professors, say, who could boycott because a fellow teacher had been dropped from the school for a belief he held. No indeed, – a president might earn our votes, but our respect was reserved for the noble novelist who spoke out for what he believed, and we grew up believing that what very little we could do to make our country more to our liking would not be done through direct action (politics) but through indirect hinting (art of some sort) or, if we were lucky, we might have chance to take part in some public demonstration or other.

How did it happen? How as it that this careless vote became the thing of which most of us will be proudest in our lives? How did he win our respect, which we so carefully guarded, and which we had forbidden any other person in the "field" of public affairs?

What Mr. Kennedy did for me was to show me that a good mind <u>could</u> find a way, through direct action, to make things better. At first I didn't believe it. I couldn't. But then I did. And no professor's speech, no writer's "letter to the editor" could compare with the easy way your husband won my faith and changed my thinking, so that from now on I will watch politics with a critical eye, and I myself will not rest until I find a way, however small, to change this country and make it better.

Horror, confusion, and madness are things we all fear and wish to protect our children from. Why, my mother tried to protect us from the horror of my father's death when I was ten. Now she regrets having tried, for we felt it anyway, and we were ready to mourn and to weep. But she was afraid for us and wanted to shelter us from horror. And death is horrible – some deaths more horrible than others. All human beings experience the violent disruption caused by death, and to try to coat it with dignity is not right.

I will remember the horror of his death the rest of my life. I will remember his greatness and what he meant to me all my life. The two will live side by side. Neither shall blot out the other. Time will heal the pain of both, but I shall remember them both equally. That is life – horror and greatness, and in order to achieve balance we must keep both in mind.

Forgive the severity of this tone, and please accept my most humble sympathy. Words cannot express the depth and quality of the love my friends, my generation and myself feel for both you and the very vivid memory of your husband, whose light shall indeed burn forever in our hearts.

Sincerely yours,

Ellen Diamond

BETHESDA, MARYLAND
NOV. 23, 1963

Dear Mrs. Kennedy,

I am not a writer of letters to anyone. I am a scientist, a physicist, older than you and younger than Jack Kennedy (which is how he was called in my thoughts although we never met). Born into a depression, trained into a terrible war, and made cynical by the cold war abroad and a McCarthy at home—I found for the first time, in Jack Kennedy, a man to believe in.

As a very minor worker in the government, as a member of scientific advisory panels, for once I felt there was something to work for—and someone to lead whom I could trust and respect. May I presume to offer my personal sympathies, and to assure you that there are many of us all over this country who hope that in some small way we can carry on the work he started. I, for one, intend to try.

To make up for the words I cannot write, may I share with you the

words of Stephen Spender that have run through my head for the last
two days:

> "I think continually of those who were truly great. . . .
> The names of those who in their lives fought for life
> Who wore in their hearts the fire's center.
> Born of the sun they travelled a short while toward the sun,
> And left the vivid air signed with their honor."

Sadly,
John Steinhart

···❋

ROCHESTER, NEW YORK
DECEMBER 4, 1963

Dear Mrs. Kennedy,

We were graduated from Rutgers University in 1959, a year that
CBS Television chose to make a documentary at Rutgers about college
students—called "Generation Without A Cause". That title was apt. We
knew what was right and what we believed in, but we had no one rallying
point, and no leader to put our convictions into words.

And then John Kennedy ran for the Presidency of the United States.
What a difference it made to all of us!

When the Berlin crisis erupted, my husband was at summer camp
with the New Jersey National Guard. I was very afraid that he would be
shipped overseas before I could see him again; yet, mentally, I was attending to details and girding myself for the strength I knew I would need.
The President was a mainstay of that strength. He told us that the country needed those men, and that everyone would have to sacrifice, and so
we were ready to do so.

As his term progressed, problem after problem came up and was met
head-on—civil rights, and Cuba—even though we didn't talk about him
as much, we knew he was there, and we believed in him. He embodied
everything we could have asked for in a President.

Today, words are inadequate to express the loss we feel. That awful Friday night, when we were talked out, and drained, and trying to sleep, John said only "Our brave young leader is dead". But even though the enthusiasm and fervor the President inspired in us will never be quite the same, the dedication he inspired will. And your great strength has been an example to us.

We just wanted you to know how much we loved him.

Sincerely,
Lynne and John Clarke
(Mr. and Mrs. John W. Clarke)

[JANUARY 1964 POSTMARK]
PHILADELPHIA, PA

Dear Mrs. Kennedy,

I hesitated in writing this because I felt that any words I could express would be meaningless. But then I remembered something that might express what I haven't before been able to say.

In 1960 I was in my second year of high school. Politics never interested me but, the then, Sen. Kennedy aroused my curiosity. His straightforward manner made me sit up and take notice. I found myself reading his biography and his books. I also began to read sections of the morning paper which I never even glanced at before. By the fall of 1960 I knew a good bit about the election issues and found myself arguing things like : Quemoy and Matsu and tax cuts and other things. I must admit, I managed to swing a few votes over to the Kennedy platform. The only reason being the fact that I always made sure my opponents didn't know a thing about why they were voting for whom they were voting. Anyway, to get to the point, Sen. Kennedy's personality acted like a much needed and awaited booster shot to me. And as we saw, later, he acted as a booster for our whole country. Without even realizing it, my marks began to improve and by November, I was on the honor roll.

I've never told anyone this before. My parents could never understand my new-found energy. I never told them because I felt it to be a personal thing between the President and myself. Later he inspired all the youth of this country that way. He was the right man at the right time.

I've saved all the papers with the stories of that day. Perhaps, for my children so that I can show them the headlines of the end of a good thing . . . a good thing that isn't to be forgotten. Perhaps I shall say to them the lines of a song which you mentioned, only adding a few lines more: "Each evening from Dec. to Dec. before you drift to sleep upon your cot think back on all the tales that you remember of Camelot."

I know that song well. I always became sad at the end but could never fully understand why. Now, after seeing a man who held the dreams of Arthur and the hopes, I now know why I grew sad.

I hope you know what I'm talking about. In Life you quoted another line from the song. I hope you did. If not you probably don't understand a word of this. Everything was just so confused in those few days that perhaps the magazine was wrong.

I'll close with one last thing.

In late October of 1960 Sen. Kennedy came to Philadelphia and was to come down a street near me but because of an arm injury while shaking hands, he was unable to drive by his planned route. I remember standing for three bitter cold hours and never even caught a glimpse of my beloved candidate. And then, remember standing, on October 31st in the cold waiting for a glimpse of a President, a great President. That day, my dormant wish of almost exactly three years came true. I saw a slender, magnanimous man emerge from a blue convertible before entering the Bellevue-Stratford Hotel. The glimpse lasted a few seconds—the impression will last a lifetime. As will the love this country had for him. It will last more than a lifetime, for his story is an American Heritage to be passed down from generation to generation.

There is nothing more for me to say. You've received letters of sympathy from all over the world, with their grammar more eloquent, and the writing much neater. But, believe me Mrs Kennedy, the thoughts could not be more sincere.

I'll miss him so, We all will,

Sincerely

Janis M. Lockeby

ST. OLAF COLLEGE
NORTHFIELD, MINNESOTA
NOVEMBER 26, 1963

Dear Mrs. Kennedy and the Kennedy family,

In your moment of supreme sorrow, I would like to express my sincere sympathy and, unofficially, that of the students of St. Olaf. Just as the tragic news reverberated around the world, it sharply penetrated into the indifference and apathy so often found on college campuses. For most of us, the news of President Kennedy's death was a personal tragedy, one that shook us to the very roots of our being. As the campus community gathered in the Chapel on Friday night to pray for comfort for your family, for our own sustenance, and for strength to face the uncertain future, we found some consolation in the beautiful poetry of the one hundred and twenty-first psalm "I will lift up mine eyes unto the hills from whence cometh my help. My help cometh from the Lord, which made heaven and earth."

I only saw the President once—as he descended from his airplane in Rockford, Illinois, during the 1960 campaign. But even though I had only this brief personal contact, his assassination destroyed part of me as well as part of most of us. I suppose we, as students, identified ourselves with the exuberance, the love of living, the ambition, the liberal philosophy, and the belief in the goodness of mankind of our youngest chief executive. We were born during the war that tempered your generation, grew up during the harsh and bitter peace of the bipolar Cold War, and still have not achieved full maturity. We identified ourselves with President Kennedy because his ideals are ours, and we admired the man who displayed so much courage, conviction, and faith in man. Our idealism has once again been shattered in the face of human depravity, and it has left an emptiness which will not soon diminish. But we have also been left with an inspiration in the life of your husband, father, brother, and son.

This probably sounds insincere or eulogistic, but this letter is not meant to be so. Through this tragedy we as students, may have found what we are forever searching—direction for our lives. We hope that President Kennedy's faith in America will be justified in the future, and that, eventually, it will be our generation that upholds this trust without flinching. In his death he became a martyr for the cause of peace, justice, freedom and brotherhood. May God bless you all.

Sincerely yours,
Gretchen Lundstrom

POMPANO BEACH FLA
DEC. 11TH 1963

Dear Mrs. Kennedy

I hope this will be a little different than the many letters you are receiving. If it does not get to you through your screening committee at least I've tried. Had your husband lived I could have someday gotten this message to him. I feel I should try now with you.

I thought often of sending my story to the Readers Digest or its like. I think it's a story of an unforgettable character. Here goes

I am a father of 3 girls—now adult. The first two, lovely girls, pretty well got through the morass of this modern freedom unscathed. They were, all through school honor students leaders in activities, in short good at everything.

Then the youngest. She was shy, gauche and as she grew became more withdrawn. My wife and I knew that at every turn, She was meeting tales of the ability of her older sisters. In spite of our efforts to establish her as an individual, She with drew more and more. She was talented in many ways different from her older sisters but in off beat things. Since She was in parochial school, the good nuns were always telling her of her older sisters' accomplishments. Seldom did she speak to me, other than routine.

Suddenly, She started to become interested in current events. She would take me aside and tell me of my responsibilities as a citizen. I encouraged her, but since our family always voted , and carefully, I did not pay too much attention. But, She kept after me—then my wife & then our neighbors.

I started to get requests from this child for books, magazines, etc. Naturally, we encouraged her. Then she started to buttonhole my friends and spoke to them of their duties to country. I got interested and questioned her.

It seems her interest was centered in a Young Congressman I read some of her stuff & I got interested.

I'll never forget the convention at which John F. Kennedy almost became Vice President. I have never seen eye to eye with Adlai Stevenson or in fact any of the Harvard ilk like Alger Hiss et al. But the daughter kept pressing. When he was finally defeated, a dark pall came on our daughter.

Meanwhile, our older daughters and our friends noticed the change in our youngest. We all encouraged her, but She now became disheartened. Then, She found He was going to run for President. She started to campaign for Mr. Kennedy, even tho' still too young to vote.

Well the campaign your husband waged, his election and the end are now history. The daughter was truly a part of all of it.

During your husbands administration we, now transferred from Pennsylvania to Florida, received a day by day account of the President's activity. Even during the Cuban crisis, we in Florida, were, understandably worried. She staunchly insisted the Presidents way was right.

In closing, this girl, through her study of Jack Kennedy, all through his career, became mature, alive, tremendously informed of her heritage. She gained courage to the extent that she is now away from home—on her own, working at Atlanta, Georgia, and doing well.

We know, by her letters that she prays for him daily. We also know that John F. Kennedy was heard at Inaugural by one person at least "What will you do for your country."

[This requires no answer]

Respectfully
Henry H. Delaney

USNH BEAUFORT, S. C.
23, NOV 1963

Dear Mrs. Kennedy,

I'm a young Marine recently injured in training and now I'm in the hospital. I hope you'll read this, it will take alot off my chest because I have a guilt that I must confess. Your husband meant something to me even though I didn't realize it until his death.

I remember in my last year of high school how disappointed I was that the President had not taken action on Cuba—then he did just that. I had many an arguement in my classes about the topic and lost alot of friends. I disliked him because he took so long to act. Then trouble erupted when he took action. Ma'me he made a fool of me when I heard how he had such heated debates with officials and still kept them as friends. I pray to God that some day I'll be able to do the same.

I really don't know why I'm writing, but I was reading the Bible and it was as if God had spoken to me. You see my Mother passed away a few years ago and it brought me closer to God and as I was reading in the Bible tonight I came apond this verse from Luke 12:4-8 "I tell you, my friends, do not fear those who kill the body, and after that have no more that they can do. But I will warn you whom to fear: fear him who, after he has killed, has power to cast into hell; yes, I tell you, fear him! Are not five sparrows sold for two pennies? And not one of them is forgotten before God. Why, even the hairs of your head are all numbered. Fear not; you are more value than many sparrows."

Mrs. Kennedy your husband was a great man and his words were first with me when I was in boot camp. I came into the service hoping to do my part and that's when I found those enspiring words over our squad bay hatch" Ask not what your country can do for you, but what you can do for your country". These words are over every squad-bay hatch on Parris Island.

When I saw your husband portrayed in "PT 109" I couldn't help but like him

Mrs. Kennedy I'm praying for you and our family. I hope you'll forgive me, all Marines aren't bad, please forgive me.

God Bless you and Mr. President,

Sincerely,

Pvt. Robert W. Zemeski

N.Y. N.Y.
[FEBRUARY 3, 1964, POSTMARK]

My Dear Mrs. Kennedy

In 1957, when I was a freshman at the University of California, my roommate and I put up a sign in our room which read, "Kennedy for Emperor!" And we meant it; we cajoled democrats: we tyrannized republicans.

You might imagine how delighted I was to attain my majority on Nov. 4, 1961, just in time to vote for him on Nov. 6. It was the best present I could have had.

Election night was unbelievable; those of us who managed to stay awake were sorely tempted to burn down the Oakland Tribune building in celebration.

The effect of his election and the force of his personality was tremendous at Cal. Students who had long been disillusioned with politics and discounted it as an ignoble profession, who considered it a kind of legalized crime, and those who practiced it opportunists, suddenly changed their minds. Because, at last, there was a Man in the White House, a Man who had the courage to lead, to <u>really</u> lead, to serve this country the way <u>he</u> thought it should be served. There were so many reasons to love him, his unjaded approach to the world, his new ways of getting things going, of doing what was <u>right</u> without too much attention to the ballot box. This is what we had always wanted from the President, and at last we were getting it.

We loved him very much.

I think we all proved it one May, when the University was celebrating its anniversary and your husband was coming to speak to us. We had to use the football stadium and it was full an hour and a half before he arrived. When he did, the whole student body, possibly one of the most unsentimental, cynical, even unaffectionate groups in the country, stood up and applauded until their hands hurt—many cried. It was wonderful to be so very proud.

Most of us called him "our leader" but we didn't feel it was disrespectful; after all, he was, as no one else had ever been.

November 22 and the days that followed were the most miserable of my life. Things can never be as good again, for me or for anyone else.

On Christmas Eve, I came across a line from Horace that was of some help. I thought you might like it too:

NON OMNIS MORIAR—I SHALL NOT WHOLLY PERISH.

> With great respect and admiration,
> Dona Fowler

....❧

My Dear Mrs. Kennedy:

Through the United States and all over the world, there are millions of people like me. We do not move in important social circles; we do not make decisions which influence the lives of huge numbers of our neighbors; nor do we often write letters to public figures to express our thoughts and feelings. We are seldom even visited by the pollsters who profess to know how we think.

We are, in short, the millions who have been dumbly suffering with you since the moment of the murder of your fine husband—our good and great President. Surely nothing I can say can mitigate your grief and the sorrow of your children which will grow with the years when they are more and more aware of the loss of their father. I can only do what I know many of the voiceless ones are doing. I can only pledge you that the influence which your husband had on us and our country—the world, in fact—will not be allowed to wither because we have forgotten.

Your husband was the first president to be of our generation. We have known all or part of two world wars and many smaller conflicts and we have been sick with yearning for peace. We have yearned to feel that our children and their children would grow up in a world which had given up senseless slaughter and he—he gave us hope.

The great ones of the world have spoken their condolences. Now I presume to beg you not to forget nor to let your children forget that they had a father who moved the hearts and minds of men. May your bright courage never fail you. We have watched with silent pride as you did what had to be done and did it in a way which magnifies wives everywhere.

Very truly yours,
Tom Emmitt

PENNSYLVANIA STATE
UNIVERSITY
UNIVERSITY PARK,
PENNSYLVANIA
JANUARY 14, 1964

Dear Mrs. Kennedy,

When I think of the countless number of letters you must have already received, I feel guilty about adding mine to your burden, but I feel strongly compelled to write you for whatever comfort it may bring you, and that is why this letter is in your hands at present.

It has taken all this time, to <u>begin</u> to be able to express what I have painfully felt for the last two months.

A few days before Christmas, a friend's illness forced me to leave my husband's side here at Penn. State and travel by myself to New York City to comfort him.

. . . This necessary week away from my husband reminded me once more what loneliness means, and gave me an idea of what you must feel. I think that knowing, loving, and living with a dynamic person and then losing him, must be much more painful than the loneliness of never having loved.

Christmas Eve, I looked at a photograph of my husband, and thought to myself, "what if this were all I had left?" I am beginning to understand.

Your husband made an unprecedented impact upon people of all ages. I think his main contribution, in his life and in his death, was to give America a conscience. I am proud that I respected your husband's ideals throughout his short term as President, and that it did not take his death to make me appreciate what he had accomplished and the unfinished things he wanted to accomplish. One of my biggest privileges was going to be able to vote for your husband in the 1964 election. (I was only 19 years old in 1960.) But, his ideals have taught me to give of myself, in any way I can, for my country, even if this cannot include casting a ballot for the most truly magnificent <u>leader </u>this country has ever had. I say <u>leader</u>, because although we have had many presidents, we have had few true leaders.

New York is a place that makes you feel you're on the top of the world, or a lone being in a stream of humanity at the bottom of a concrete chasm. I have felt—and still feel—the latter. Thoughts enter, cross, and recross my mind: that one man should have had the power to, in such a short instant, cancel out all the things that the man <u>was</u>, as well as the man himself; the following days of mourning when people demonstrated a long-lost decency to each other, and even those who couldn't express their thoughts, began to think; the blossoming of JFK records, pictures, pens, etc., by those who wanted to give the common man something to hold onto, as well as those who were lecherously capitalizing upon the occasion, perhaps creating a bigger sin and demonstrating a greater sickness than the man who killed our President.

This has been the legacy: good or bad, it stands.

My husband and I had applied for Peace Corps duty before the tragic event of two months ago, and, ironically, we were notified of our acceptance as trainees on November 22, 1963. We are more determined than ever to do our job in the Peace Corps, and, speaking for myself, more strongly, as an American citizen than before. After the Peace Corps, we will also be very conscious of our contributions to our city, the nation, and the world. Your husband is one of the guiding influences we have had in becoming responsible citizens of this nation and the world. The debt I owe to your husband is carrying out what he has taught me: <u>caring</u> is not enough, <u>doing</u> is what counts. This is why service in the Peace Corps is so important to my husband and me. President Kennedy gave us the opportunity, it is our responsibility to take advantage of it.

I truly hope that this letter serves as some sort of comfort to you, even it it is small; that there are two people here who are aware of the man that was, and the ideals that were the man.

Karen S. Kendler

....✺

CAPTAIN PETER NEAL O'CONNOR
UNITED STATES AIR FORCE
GRIFFISS AIR FORCE BASE
NEW YORK
NOVEMBER 25, 1963

Dear Mrs. Kennedy,

Please accept the deep sympathy and prayers of Mrs. O'Connor and myself for you and the children in this, your time of need. Words can not convey our grief over the loss of our monumental leader and beloved President. Your courage has been our strength.

I leave shortly for Viet Nam. His spirit and dedication go with me. Although we never met John Fitzgerald Kennedy, we knew him well and loved him deeply.

Peter N. O'Connor

Dear Mrs. Kennedy:

You will probably never read this letter. Certainly my sorrow can in no way alleviate yours. But with the people of this nation, I mourn the death of a man among men; and however small this tribute, it is all I have to offer.

For on this day of Thanksgiving, I find myself remembering...

In Guadalajara, Mexico, I wait impatiently for the bus under a molten, blazing sun. A street sweeper approaches hesitantly, and with a "perdóneme, señorita," says "Whom do you wish to be your president?" I am ashamed that I cannot answer, and at my uncertain response, he tells me of one John Kennedy, the senator from Massachusetts who has just received the democratic presidential nomination. The scene is repeated many times that summer, and when I return to the United States, listen carefully to the words of that young senator. I remember...

that I am not quite old enough to vote. But I am young enough, that election eve, to trudge from house to house in sub-zero Minnesota weather to speak to others of John F. Kennedy. And I remember...

Two days later I awake with the question I dare not ask. To my unspoken plea, my roommate says, "He won." A telegram goes out that night: "Two democrats in Lutheran Republican college send congratulations to our President."

In the days and months that follow ... the Berlin wall—the Cuban crisis—civil rights—the wheat deal. President and Mrs. Kennedy ... in

Hyannisport—at church—with their children. The nation has found its voice, and that voice rings out to the "new generation of Americans". Our natonal pride is renewed; our faith in God is strengthened. We lift our heads and proudly scorn the prophets of doom, the voices of the lost generation. John Kennedy is more than a president. He is the image of America itself. And then...

What was is suddenly no longer. I enter my classroom to say—what? to those anxious voices that ask, "Is it true?" They stand for prayer, but I have no voice to pray. The room is silent, the English lesson forgotten. And then, the final, irrevocable word comes through. The President is dead. Their eyes beg me to deny, but I have no voice. He is buried three days later, and with him is buried a part of every American.

To you in life, Mrs. Kennedy, go our prayers and our deepest sympathy. To your late husband in death goes our pledge—that we will carry the torch that he lit—and that the world will be lit in his name. For the watchman did not wake in vain.

With heartfelt sympathy—

Marcia Schwen

NOVEMBER 23, 1963

Dear Caroline and John,

This may well be the most selfish letter that was ever written but unless I spill out the thoughts that are milling thru my mind this grey day in November, I am not sure I will ever be able to justify future words of encouragement that may be my lot to utter to anyone.

We have children your ages and I wonder what we will tell them in defense of a society that could spawn the type of mentality which uprooted your young lives, when they read in history of yesterday's episode?

In defense of those among us, your fathers contemporaries, who looked to him for leadership and courage, may I humbly say that we are left as bereft of spirit as your family is this day.

When John F. Kennedy was elected president, I did not stand among those who had voted for him. Why? Because I was afraid—not for his physical well being (this never occurred to me.) Not because I doubted he could do the job, but because I was anxious about what the Job would do to him. He was a Christian idealist—I had misgivings about his stepping into such a secular market place, for the pressures which would be brought to bear upon him. "Better he should suffer the loss of the whole world, than that of his soul." Somehow I could not bring myself to contribute such a burden upon such an outstanding Christopher.

After he was elected, I joined the throngs in sending congratulations, and even reminding him of the above and pledging my daily prayers. I never forgot them for a day, and somehow, I do not believe he forgot them for an hour—I am no historian or student of politics, but somehow I feel that in the dramatic hours of decision which faced him from time to time (and they were many in this decade which he began) God himself must have been looking over his shoulder.

As a mother of a one, and three, and five year old, I lived with a comfortable feeling of having our specific world being currently guided by a father of a two and a five year old—feeling that he was basically working on the kind of daily course of humanities which he hoped to present to you—his heirs. He could not bequeath it to you without doing so to us and so, I lived with a comfortable feeling.

When you are old enough to read this, I hope you will be understanding enough to realize that many, many of us felt a loss equally as acute as yours, and maybe even more difficult to bear because it had to be done in muteness.

Please God, may each of your worlds be as good as your father's high hopes and may each of you grow in grace and wisdom (and forgiveness) so that his life will not have been in vain.

Your mother may well wonder how to explain us, the citizens of the sixties, to you, when you grow to a knowledgeable age. This is how I will try to tell such a bitter page in history to our children. We needed your dad to lift us from the tired, trite, sophist era of politics and politicians. We needed him to remind us all that it could be fashionable to be good and aspiring, strong and witty, determined and unbending, hopeful and charitable—in varying degrees as the need demanded. We needed to be reminded. He did this. I hope time will prove that enough of us have continued to remember his high hopes for humanity, and individually we have considered the effect our actions have on those about us. We shall certainly be grateful, with history, for having known such a president.

Most sincerely,
Midred F. Hesch
(Mrs.) Robert L. Hesch

JANUARY 16, 1964

Dear Mrs. Kennedy,

Fifty-five unbelievable days have passed since that dastardly act was performed that took the life of your husband, our president.

For the majority of us it has been a mechanical existence, never forgetting all the deep grief and pain you are undergoing. As you stated on your heart breaking telecast "All that bright light gone from the world."

Speaking for my family I can say a part of our hearts are buried with him.

I can remember standing for four hours in the rain back in 1961, when his motorcade was going through our town. When the motorcade finally passed it was worth every second of the wait. Our youngest son was then 3 mo. old, he was kept dry under my husband's coat. We both were so certain at the time that John F Kennedy would be the only president he would remember, as we did Pres. Roosevelt. On another occasion—July 4th, when Pres. Kennedy honored Philadelphia by appearing at Independence Hall—My husband and our seven children and I went down town again to get another glimpse of your wonderful husband. This time we saw him in the sun, a sight we all hold sacred in our memories.

In my own husband's words "I haven't felt a sorrow so great since my father's death.

It may be five or ten years from now that this letter will be in your hands, but the feelings expressed will be the same. Surely this generation has a deep scar on our hearts which we will carry to our graves.

Your cross is very heavy. I wish we all could help you bear it, but I know words are such an inadequate means, yet our only way. May God bless and help you, Caroline and Jon-Jon.

Lovingly,
Mrs. Paul F. Smith

GRIEF AND LOSS ·································

"The Burden of His Death"

Okla. City Okla.
Feb. 14 1964

Mrs Kennedy;

I'm just and old 73 year old man who lost his wife in 1963 and I can feel the sorrow you are going thru. My wife died in my arms as your husband died in your arms and when I watched you on Television as you walked behind that flag draped Coffin, I cried my eyes out.

I know you received thousand of letters, but I just wanted to tell you I think you are one of the most wonderfull women on earth

May God always bless you

Sincerly
R.R. Louk
1125 N.W. 29
Okla. City, Okla

The four days of televised coverage after President Kennedy's death in November 1963 constituted for Americans a unique experience. The country had grieved for and buried other Presidents, of course. But the shared and simultaneous national mourning that was made possible by television on this tragic occasion was unprecedented. Still, after thirty days the formal period of national mourning ended. And the American people, like a bereaved family, went back to their lives and forward into the 1960s, leaving John F. Kennedy to the pages of history. He would live on in memory, but the shape of those memories would shift and turn along with the country itself over the ensuing nearly half century.

Meanwhile, a private but sharp sense of loss persisted among many Americans for some time after the assassination. It is hard to recall today that the culture of self-revelation and public confession that is so much a part of contemporary America did not exist in that period. The early 1960s, in this respect, seems closer to the Victorian era than to modern America. The world of manners then stressed propriety, decorum, and deference. Many considered rectitude, reserve, and reticence as virtues rather than regrettable vestiges of repression one ought to strive to overcome. In 1963 public outpourings of grief and mourning were uncommon—reserved for funeral rites and Memorial Day. The wreaths and flowers laid in Dealey Plaza to denote the spot where President Kennedy was assassinated marked a rare, spontaneous collective grief response. More common were private expressions of sorrow, shared in conversation with friends, coworkers, priests, ministers, rabbis, family members.

In this sense, the letters many Americans sent to Jacqueline Kennedy seem paradoxical—a fact the writers themselves were acutely aware of at the time. For they revealed deep emotion to a public figure with whom they lacked a close personal relationship. Many began their letters by mentioning the contradiction, admitting they felt awkward in writing. Some politely noted Mrs. Kennedy's enormous volume of mail, disclaiming any wish to increase her burdens. Others puzzled over their impulse to reveal themselves to the First Lady, given, as one man from Alabama noted, "that I am a stranger to you." "The fact of the matter," a Californian pointed out, "is that both you and your late husband grew very dear to us (me and my family) during the all too few years we had to get acquainted with you through the media of television, radio, magazines, and newspapers. Yet I feel that I knew you both very well." Another young wife confessed, "At first, I thought it wouldn't be right for me to write to you since you never heard of me or my family . . . I started thinking of the many things we had shared with you in these brief years . . . This is my reason for writing to you, even though we aren't important except, We are Americans."

Expressions of grief and loss inevitably form the core of condolence letters, whoever the bereaved and correspondent. The letters written after the assassination of President Kennedy are no exception. Those who had suffered the premature loss of a husband or a parent particularly identified with Mrs. Kennedy. But in conveying the nature of their loss, their sense of connection to JFK, and their desire to console his widow, Americans also revealed much about how they perceived their own lives and the historical upheaval they had experienced. "I feel his loss just like a mother to her child," wrote an eighty-six-year-old Ohioan. But there was also this, she added: the President had died for his country by "a most disgracefull hand." Those who had endured violent and catastrophic losses saw their own tragedies revisited. One young mother who lost her husband and four out of five young children in a fire wrote several months after Kennedy's assassination: "Our lives and way of living are very different—still we have suffered the same losses." She recalled the letters she had received from "relatives, friends and strangers" who were "all anxious to help in every way they could." It seems, she concluded, "no matter what we lose . . . someone has lost more." Some of those who sought to comfort Mrs. Kennedy confessed a sense of helplessness and guilt—echoing a common response to death that was magnified, in this instance, by both the distance and closeness many felt to the slain President. "It is terribly difficult for this American to try to console you on the death of our President," admitted one woman "because I am still unconsolable." Like many, this writer felt implicated in the assassination. "Mine was one of the many votes," she noted, "that helped speed him to his ghastly death."

The texture of Americans' grief and loss in the aftermath of November 22 reflects, then, familiar human responses to bereavement. But it also gives evidence of the extraordinary historical realities that had deprived the country of its President. This final collection of letters reveals the eloquence, sincerity, and generosity of spirit that characterized much of the American response to the death of President Kennedy. In trying to come to terms with his assassination, many citizens drew upon their own deep reservoirs of life experience to find meaning and affirmation in a tragedy they felt had changed their own lives as well as the history of the nation. Writing the letters itself provided some catharsis. "Maybe I didn't do the proper thing writting to a great lady like you," one African American housewife in Chicago fretted. "I had to take the chance. You see sometime you start hurting and can't seem to stop."

No group among those who wrote condolence letters identified more with Mrs. Kennedy than those who had lost a spouse—the majority of them widows. The following messages, arranged chronologically, come from those who saw in the untimely death of John F. Kennedy a loss with which they felt all too familiar. They offered solace, reassurance, and often praise to Mrs. Kennedy in equal measure as they attempted to draw upon their own hard-won wisdom in comforting the former First Lady. ❧

My dear Mrs. Kennedy,

With full realization that this will be but one of hundreds of expressions that you will receive, and the possibility you may not personally read this, I feel compelled to compose this letter to you.

It was only last May that my children and myself received a letter of sympathy sent by you and your husband. My husband, who was 26, had been killed while on a routine Air Force mission as an electronic warfare officer. Suddenly, without warning or realistic preparation I found myself alone and my children—Jackie Lynn, 4½, Rodney, 3½, and Bradley, 1,—fatherless. We had been married five years; two of which were spent in college, three more in advanced training in the Air Force. We had lost our third child shortly after a premature birth. Just as we seemed on the threshold of rewards of preparation, it was ended.

Now I must write this letter to you and extend my sympathy to you and your children. But it is more than sympathy; I extend to you deep and sincere empathy. I know nothing that I nor anyone can now say has any meaning or comfort for you. I know the anguish and heartache your whole being is experiencing, and the bewilderment and disbelief that yet crowds your mind. Senseless acts! Why do they happen?

I have no words of wisdom nor no words of comfort. All I can tell you is for the first time since my husband's death can I feel for someone else. Though unimportant and of modest means, I feel so akin to you and wish that I could put out my arms to you and embrace you and somehow help. I pray that the traditional and trite comments from those around you will not hurt you too much. People who care somehow never know really how to express their love in a meaningful way. I pray that the national aspect involved will not disgust and embitter you. The whole world has truly lost.

No more can I say, only to repeat myself. Please accept my inad-
equately expressed emotions. My whole self goes out to you.

Sincerely,

Sue Ann Andersen

(Mrs. Einer E.)

SPRINGFIELD, ILLINOIS
NOV 23, 1963

Dear Mrs. Kennedy:

Here I sit at 4 a.m crying my heart out. Just how you feel at this time
I know and words cannot express my feelings. On Sept. 2, 1963—Labor
Day a tragedy such as this came to me—my husband drowned at our
Lake Springfield so I am sharing your sorrow more than anyone but God
knows.

I know you will never see the letter as there will be so many it would
be impossible for you to read them all but maybe somehow you will hear
about it.

I'm not a catholic but my prayers for you & your family are sincere.
Have faith and the Good Lord will truly see you thru this ordeal—I know.
It's not easy—believe me. Please excuse the looks of this paper but the tears
won't stop & I feel like I have to write this even tho you'll never see it.

As I go to the cemetery this morning which I do every Sat or Sun my
thoughts & prayers will be with you & your family also.

I find so much comfort in the words of John 14, 1-6 and the 23rd Psalm.

Please accept our expression of sympathy in our poor but truly sincere way.

Sincerely & sorrowfully,

Mrs. Elizabeth Pucka & Becky

P.S. I heard over the radio that the President's casket is bronze. We buried
our poor Leo in a bronze casket too.

TELEGRAM

LSC992 NSA554

DA351 D LLN364 PD FAX DALLAS TEX 24 608P CST

MRS JOHN F KENNEDY

WASHDC

MAY I ADD MY SYMPATHY TO THAT OF PEOPLE ALL OVER THE
WORLD. MY PERSONAL LOSS IN THIS GREAT TRAGEDY PREPARES ME
TO SYMPATHIZE MORE DEEPLY WITH YOU

MRS J D TIPPIT DALLAS TEX

(34).

EDITOR'S NOTE: *Mrs. Tippit's husband, J. D. Tippit, was a Dallas police
officer who was murdered by Lee Harvey Oswald on November 22, 1963,
forty-five minutes after the assassination of President Kennedy. Officer Tippit, having heard a police description of the suspect radioed to his cruiser,
attempted to detain Oswald who was on foot. Oswald pulled out a handgun
and shot Tippit point-blank four times in broad daylight, in front of several
witnesses.*

NOVEMBER 24TH, 1963
WASHINGTON, D.C.

Dear Mrs. Kennedy:

I write you not as an important person but as a widow who was left
twenty years ago, after twenty-two years of blissful and exciting marriage,
with two small children. And I want to say that you must not fear for
them. With such a heritage of courage and conviction, and enshrined in
the hearts of the American people as they are, they will go far.

And for you, with a bleakness ahead that the presence of princes and emperors can do nothing to assuage, you are relieved of the one intolerable burden when we lose someone we love, remorse. You made your husband happy—and as an American I thank God that both of you were allowed to lead our nation even these few years.

With deepfelt sympathy,
Katharine Stanley-Brown

SAN ANTONIO, TEXAS
NOVEMBER 29, 1963

Dear Mrs. Kennedy,

You and President Kennedy were in my office a week ago yesterday.

I am secretary to General Bedwell at Brooks Air Force Base, and I will forever be haunted by how handsome and healthy and happy you two looked—and how gracious you were to me. We loved you here at Brooks and feel your loss especially deeply.

You have every reason to be consumed with bitterness, but I hope you can find it in your heart not to be. I have learned through the years that you can live with sadness but you cannot live with bitterness. It destroys you and those dear to you. My husband, a B-29 pilot, was killed during World War II when I was three months pregnant with our first baby.

Everyone marvels at your courage. You are indeed worthy of being a President's wife—and President Kennedy's wife in particular. You make us all proud to be Americans.

May God be with you, my dear Mrs. Kennedy.

Very Sincerely,
Claudine R. Skeats
(Mrs. A. E. Skeats)

Dear Mrs. Kennedy,

You have my deepest sympathy on the loss of your husband and also your baby.

I wanted to write you sooner but I couldent get myself composed. The reason for my writing you is that my husband took a heart attack and died instantly a few mi. after your husband was lowered in his grave. My husband was also 46 and served in the Navy during the second world war. Honorably discharged.

My children (I have 2 a boy & a girl) & I are having a difficult time convincing ourselves that he is gone.

My husband felt so bad when President Kennedy was shot. He was terribly upset, and while we were watching the funeral on television he took this heart attack & died.

I don't know why Im even telling you all this, you probably will never read my letter, but anyway when I lost my husband right after you lost yours I felt close to you and I wanted to extend my deepest sympathy to you. I understand and know the loss you are feeling, because I am going through the same thing.

May God Bless you & your children.

Please pray for us we need the courage to go on. I will pray for you and your children.

sincerely,

Mrs. Margaret Mesaros

PITTSBURGH, PA.
NOV 27, 1963

My Dear Mrs Kennedy.

I would like to express my deepest sympath to you and your children. This was a great shock to all of us.

Just like you I lost my husband 7 years ago. He was killed in a steel mill. I was left with 4 children. My husband at the time was 27 years old. My twins were 10 months old at the time. So believe me, I know what you are going throught. My heart go out to you and your children.

The day that Mr Kennedy was Assassination my children came home from school crying, and its been six days now and we still are crying. He was a wonderful man, and a good father to you children, he also was the best President we have ever had. We will never forget him. We will remember you and your family in our pray's.

My God Bless you and your children.

Mrs Joseph Bradley Jr.

HOUSTON, TEXAS
DEC. 2, 1963

Dear Mrs Kennedy,

I am a widow of 29 and have four children, ages 11 to 5. I want to tell you how much admiration I felt for your husband. He was the first president we have had that I felt had a touch of greatness. I am so glad that I was able thru T.V. to see the services for him. I have Cancer and was too ill to go see you when you were here, but my children saw you. My boys were much impressed by your husband, but my daughter just wanted to see "Jackie." Kelly and Caroline are the same age and we read all we can about your daughter. I hope things go well with you and your children. Your loss is great but our loss is so much greater. I was very proud of our first lady and feel that the president would have been also.

Please remember me in your prayers as I have not much longer. I pray for you and yours nightly. I wish I could think of something to say to tell you of my grief for our President and the pride I feel in you but cannot.

God Keep You,

Jo Hebb

··· ✳

Dear Mrs. Kennedy,

My deepest sympathy on the tragic loss of your husband. If I could take away the sorrow and pain, I would, but I can only share it.

On April 25, 1963 I received a letter from your husband expressing his sympathy on the loss of my husband, an officer aboard the U.S.S. Thresher, which sank April 10, 1963.

May I quote a few lines from that letter, as I am sure you will find consolation from his words. "It is a sad fact of history that this price of freedom must be paid again and again, by our best young men in each generation. Your husband has joined the other defenders of this Nation who have given their lives for their country."—My dear Mrs. Kennedy, this truly is your husband.

I pray for you Mrs. Kennedy, and our husbands, and Mrs. Kennedy, when you are very much alone with only your thoughts, please, please think of us, the wives the Thresher left behind. Our hands reach out for yours in those moments of darkness.

Death is incomprehensible, a mystery, and a Divine one at that, not to be understood, only to be accepted.

Most sincerely,

Mrs. John J. Wiley

EDITOR'S NOTE: *A nuclear powered submarine, the USS Thresher was lost at sea off the coast of Cape Cod on April 10, 1963. The entire crew of 129 died in the accident.*

DEC. 1963
WASH., PA.

Dear Mrs Kennedy and family

I just wanted to write you a few lines about your Dear Husband.
I lost my Husband 15 year ago and I know how it feel for you and all
your family[.] I no it is Hard to Bear but when we half to go. I think
you have two Darling Children. I feel sory [for] them[.] when my
Husband Died we lived on a Farm so I had Leave it and I work out
By days it was allfull Hard[.] I soon be 72 years old and I am Still
workeing out. I am on the Relief[.] I only get $59 a month the way
thing is you can['t] do any better[.] if are [our] Presend John F Kendy
had of lived thing would been better[.] I talk to alot of people and
they said John F Kenndy was going to be the Best of United State[.]
I know he had to have time he would straiten things and we all could
Live Happy[.] When I heard of the President die I could Not believe
it[.] I watch the televison but made me Sick to think some one had
Kill him[.] every body said he was a good Presdent he was going to be
the Best Presdent of the United State[.] if any of his Brother Would
Run for President I would vote for them[. I am a Democrat and I am
Proud of[.] I think I will feel Better By Write to You and Your darling
children hope you will get along and May God Bless you and all you
Family and allso the President family[.] I hope you dont mind for me a
writing to you I might feel Better this a Friend of all of you
                                        Blanche Wise

RUSSELLVILLE, ARK.
DEC. 25 1963

My dear Mrs. Kennedy:-

My heart goes out to you and your little family this Christmas morning. I remember back to the year 1948 when my five children, two little girls and three older boys had our first Christmas without our beloved one. It was a sad day indeed. We were thankful tho for a good father and for the pleasant memories we had of other Christmas mornings. We had celebrated our twenty-fifth anniversary. I am sure that this morning you too have had those same thoughts and you are also thankful. My faith in eternal life was made stronger by his passing. Yours too has been made stronger. How our hearts bled for you + your wonderful family.

Be of good cheer. Jesus said, "I am come that you might have life and have it more abundantly." I believe and so do you.

My children are now grown. They have children—oh! how I love 'em. The parents are college graduates. My oldest son is with Chrysler in New Orleans. He is active in his church. He sings in the choir. My second son is an architect in Dallas. He and his family are members of Highland Park Methodist. He sings in the choir. My youngest son is a Methodist minister in Ft. Smith, Ark. The girls are married and they too are active members of our church. My youngest daughter teaches science in the Junior High school Chillicothe, Mo.

Why am I writing you about my precious family? I'll tell you why. Some day I'll read in our Arkansas Gazette that little Caroline has grown up and she will contribute service of some kind to this great country of Ours. I'll also read about little John—God Bless him. May he grow up to be the 2nd Kennedy president (maybe the 3rd) of the United States of America. By that time may all hate, strife, prejudice and dis-courteous conduct have passed away from America's way of life. May the Prince of Peace truly live in the hearts of all people.

Sincerely

Jerrine Hilliard

Dear Mrs. Kennedy,

Forgive me for writing to you—for I know that the constant opening of a wound does not help to heal it. But I feel that I owe you a deep sense of gratitude. For indeed—through your great tragedy, I have found salvation and comfort. For, you see, I too have experienced the same loss that has befallen you and in the same fashion—just seven months earlier.

On April 9th my husband kissed me goodbye and went down to his law offices and 35 minutes later he was shot and killed by a disgruntled client, for no plausible reason.

I know the world over, feels your great loss but only those who have actually experienced what you have, could possibly know what you feel deep inside.

Easter Week was my week of terror, my husband was buried on Easter Monday.

I have since, gone through, including Easter, my birthday—my childrens birthday, our 22nd wedding anniversary, Thanksgiving, Christmas, New Year's, and today my husband's 47th birthday. Each holiday was more difficult than the preceeding one, but, I'm told, you must pass through at least two sets of holiday's—to adjust yourself.

When my husband was shot, I'm afraid I was not as brave and courageous as you. My doctor feared a nervous breakdown and I lost 48 lbs. For I could not comprehend why this should happen to my dear one—who was always smiling and had such a bubbling personality. He was loved by everyone, his funeral was the largest that our little town of South River had ever known. And surely, a man so loved and respected and admired by so many couldn't have an enemy—but even our Dear Lord Jesus had his Judas.

My husband was shot at 9:20 a.m. and fought for his life until 6:30

p.m. He never regained consciousness. I have two daughters and a grand-son who was born exactly 2 months after, on the same day and the exact hour of my husband's passing. Surely, this must be God's way of replacing a life for a life.

I know that the entire world was affected by our beloved President's passing but I want you to know that my children and I actually relived all those horrible day's with you. They were so exact and similar to our own. Up until that tragic day, I was living in a very grim world by myself—shutting out my children, my new grandson and my friends. I wanted no part of the human race. And then my 16 year old daughter came home from school and said "Mother, do you know when I heard about our President, I felt the same way as when I was told about Daddy. A shock went right through me from head to toe."

And then day after day I just sat before the television and watched the entire proceedings, and slowly, very slowly, as I saw your grim face and the courage you portrayed, I felt a new strength come into my body.

Even though you showed no tears I knew better than anyone that in the privacy of your own room, you cried. For me, 9 months have gone by, and I still cry in my pillow every night.

Though I could not understand why this should happen to my husband at the peak of his career, when he had everything to live for—I felt that somewhere, somehow I would find the strength and the courage to face reality. But thus far, my depression was very great. I spent many hours with my priest and he constantly told me that God would show me the way.

And then, while watching your sweet face, day after day, I suddenly knew that God had chosen your courage and tremendous faith to show me the way. Whenever my day is bad and a little on the depressing side, I think of you, and say a Hail Mary for your husband and mine, and the day seems to be a little less depressing.

God certainly moves in mysterious ways, for suddenly, 'He' showed me the way through you, dear gracious, humble and courageous Lady.

May God help you to find peace within yourself. I know that your great love will never lessen as mine never shall, for my husband. Days shall never be the same again but time has a funny way of easing the ache in our hearts.

I know that you have received a countless number of letters and that it would be impossible for you to answer each one personally, but I would be humbly grateful if you would find time to answer me with a brief message.

May God Bless You and your children and I thank 'Him' for showing me the way through you.

> Sincerely and Humbly
> I remain
> Helen M. Milano

My Dear Mrs. Kennedy My heart has gone out to you these past weeks, you do have my Deepest sympathy, and have thought of you so much today. I lost my husband three years ago today, and you have to pass the same way before you know how the other person feels, of course you will have every thing you need financially while I was left without money. I have to sew for a living until I can draw Social Security, and have six years to go. My Husband was dying with cancer. When President Kennedy was elected, we sat up all night the night the election returns were coming in he would come to every few minutes and ask if he had won, and when he was elected he said I can die in peace now knowing he will Be President. I know they are Both in Heaven now. My heart went out to you, this week when you were on television. I know you miss him as a husband and a Father for your children. And Our Country sure misses Him as our President. I pray that God will be very near you in these days.

> Yours in Christ
> Mrs Ruth Stafford

EDITOR'S NOTE: *This undated letter was probably written on January 21, 1964—the anniversary of Mr. Stafford's death.*

KANSAS CITY, MO.
JANUARY 1964

Dear Mrs. Kennedy and Children:

Mrs. Kennedy I had to write you note. I never expect it to be answered. Because I am just a Missouri housemother. But "Bless" your heart I seen you on television thanking the world. My how I suffered over the tradedy. I didn't hardly move for 4 days. It still seems like a terrible dream. Mrs. Kennedy my husband was killed instant to. He spent three years in service. I loved and waited for him. Then he came back, was run over by a bulldozer. I was left with 5 yr. old son. Thank God wasn't for him I don't know how I would have made it. Also a handicap little girl Susan. She does everything but talk. I have spent all my money finding the cause. The Drs. tells me nothing. So I just pray. That's seems all a person can do now days. One thing I am thankful for. I saw Our Beloved President in person in Kansas City when he was campaining for to be the President. Mrs. Kennedy, would it be asking to much of you to send me a picture of you the President and your two lovely children. Altogether I have one of you, Caroline and the President. My little girl named her doll after Caroline. My Son sent in for the awards, that Pres. Kennedy said would be given to the children if their Father was killed. We rec'd ours. But my Son was disapointed. Because Pres. Johnson sign it. But death wont wait. Mrs Kennedy I think some one was asleep. When this tragic thing happen. I don't want to seem overbearing. But believe me I feel half of me is gone. I never miss one of his speeches. The whole world love him. I just wondering if the President knew. How well he was loved. Never in my life other than my family. Did a death ever effect me so.

> Well dear,
> take care, pray
> look forward
> I hope some day I can make the trip. Just to
> pass the President resting Place.
> Mary Glassner

Dear Mrs. Jacqueline Kennedy:

Months have passed since the tragic death of your husband and <u>my
president</u>. Your's was the Glory of loving and being loved—bearing his
children—to one of the greatest men of all times—surely of our day and
age. You will live with Jack Kennedy in written history as long as written
history and Earth lasts.

My husband and I sat glued to the television thru those misery filled
days of the assassination and funeral. We wept—unashamed. We wept
for <u>you</u> and <u>ourselves</u>. You did not weep in public; your courage was tremendous.

On March 11, 1964 my Jack died of emphysema. He had been ill—
with knowledge—about two years. He was not a political figure—just
a number—one—in the statistics of population and individuals of this
great country of <u>ours</u>. But he was a man with an unblemished character
and my "Glory." He was honest, good, loved by his small circle of friends
and thought highly of in a hard business field—realty. He was a broker
and had his own business. "Jack Cannon Realtor."

I lack your wonderful qualities of faith, youth, position, courage and
<u>need</u>. I roam thru an empty, two-story house filled with shadows. It is not
a pretentious home but most comfortable—and it was filled with love
when my Jack lived.

Should you plan a trip to Sacramento or vicinity, I offer you my
home as long as you cared to use it. If it is not elaborate enough
and given sufficient time, I would make whatever changes I could to
improve to your convenience and comfort. This I do offer in gratitude
for the joy of having known and lived during the times of Jack Kennedy
and his charming family. And when he died so very many of us felt a
personal loss because we loved him. He brought a closeness to his fellow citizens by sharing his problems, his family and giving of himself so

completely that we had to <u>care</u> when he died. I feel a little of us all died with him.

In my hours of grieving I feel closer to you but lack your strength. I would shout from the housetops "My Jack is gone." If I could I would immortalize his name. He shared one thing in common with your husband—<u>a love of fellowman.</u> It is reflected in the eyes and cannot be expressed by lip service. It is created by the "Way of Living" and the reflection and image readable in the Eyes for they are the windows of the Soul.

You have my deepest sympathy which I offer from my departed husband and myself. He asked me to write a long time ago but I was too busy watching him dying little by little each day, painfully and yet struggling so hard to <u>live</u>.

Please accept the offer of my home if you should ever have use of it with the spirit of love that dwell here.

Sincerely
Irene Cannon

NORWICH, OHIO
JAN 15TH 1964

Mrs John F. Kennedy,

I hardly know how to begin this letter as it is the first I have ever written you. It isn't going to be easy as I have lost the greater part of me. On the same day and the same hour of your terrible tragedy my husband dropped dead. I was on my way home from Janesville when I heard the awful news of our Dear Friend being assassinated and I was so deeply hurt I could hardly stand it and as I drew near my home I heard the sad news that my dear one had passed away. Why he was taken so suddenly from me I will never know. He was the most wonderful man to me that ever lived and I know you feel the same about your dear one. We always listened to him on T.V. and was very proud of him. He was such a bright man and why he had to suffer so in his last moments no one but God understands.

And my heart aches for those two little youngsters who were so close to their Daddy and he was so proud of them.

I know they will be well taken of but that is beside what I mean. It is not the financial part but the awful loss of their Father will linger with them for years to come. I was left in very poor circumstances and not able to work any more but we do have God's promise that he will never leave us or forsake us if we put our trust in him. My husband was B. F. Archer and we celebrated our Golden Wedding Anniversary Jan 1st 1963. He was 77 and I am 73.

We have worked hard and had lots of sorrows too. We have two lovely daughter's and one son living and one dead. They are all married and live away. I feel very much alone.

The dear Father & Mother of Mr Kennedy I know are very sad and my heart aches for them. I know Mr. J. F. Kennedy was very fond of them and his passing would be very hard on folks their age and especially the invalid Father.

We were so proud to have John F. Kennedy as President of U.S. and we did all we could to help put him there. We wish for Mr & Mrs Johnson a very good year but we know it will be hard for them as they loved Mr. Kennedy too.

Only time and God's love can help the heartache and loss and I pray God's blessing on you and your family. Now if this letter does not make sense you know my heart is broken and will never heal until we meet again.

Sincerely,
Mrs. (Pearl) B. F. Archer

Dear Jacqueline:

I know these have been trying days for you, as every time that I tuned in on the Republican Convention I had tears in my eyes and a lump in my throat. I don't know what will happen when the Democrats have their Convention. I listened to Barry Goldwater's nominating speech last night and I had never heard such a boresome speech in my life. I am certainly not a Nixon fan, but his speech was more interesting, even if he was on the other side of the fence. I thought it might have been I, but when the New York Delegates walked out of the Cow Palace and when other news commentators said the same thing that they enjoyed Nixons speech more than Barry Goldwaters.

Since you and President John F. Kennedy and the children are out of the Whitehouse, I don't have any desire to see or hear anything concerning it. In fact I don't have any desire to even vote this year. I had never been a hot Politician but I sure did change during the Kennedy Administration, but have lost all interest since you all are not in office.

I'm sending you a family picture, of you and President John F. Kennedy and the two children coming out of Church. I cut it out of a St. Aloysious Orphans Magazine, and have it framed and standing on my cedar chest and every one who sees it, wants one like it. So I contacted the Mother Superior of the Home and she sent me about a dozen copies, so I have given my friends a copy. You may have a picture like it, as not too long ago, some one showed the same one on TV. but in case you haven't I know you would cherish it as I have.

Jacqueline, I am a widow also, and I know just how you feel and what you are going through. I stayed in my home too for almost a year, when I decided to sell and return to my home town which has helped a lot, although I am twenty-five or more years older than you and don't have a family but the loneliness is just the same when your partner is gone. Although the sadness decreases as the years go on, but you will always miss him, as I know I have and it is going on four years that he is gone.

I am sending you a prayer that helped me so much. Our Faith means so much to us, so have Faith and keep praying.

Sincerely

P.S. I am sure that when you moved out of the Whitehouse thousands more like I move out with you.

Mrs. Catherine Rassche

Requests for the Mass cards prepared for John F. Kennedy's funeral, a PT 109 clip of the kind JFK often distributed, or a photograph of President Kennedy, Mrs. Kennedy, or both with their children often accompanied condolence letters. Some writers wanted a more tangible reminder of JFK.

BALTIMORE, MD.
JAN 28–1964

Dear Mrs. Kennedy:

After you have gotten your self a bit more together I'm hoping you will get around to this letter = If it's God's will I hope your heart have ease some what= we all morn with you over your lost, but we are still greatfull that <u>God</u>, left you here with the childrens (Caroline & John Jr.)

Would I be asking for to much if I would ask for a special favor? Mrs Kennedy you know; "I wish I could Own One of your HUSBAND's SUIT!"? I would Cherish it to death if you would let me have one. = I've never own a real suit ~~made of such~~ worned by such a man like Mr Kennedy I would take great care of it! And if I couldn't get a suit of his: would it be possible for me to get a picture of the family? or you & the Childrens?

<u>Thanks</u>! & Truly yours

Chris K. Gatewood

HARRISBURG PA
DEC. 5 1963

Dearest Mrs Kennedy,

I am Just Able to write you of your Sorry not only yours but All of us I never had Something to hurt me So bad after the death of my one mother in January 1963

Darling all we Can do is Pray the good Lord will lighten our burdon

Just leave every thing to him and I declair he will fix for us Our President was not only a god blessed man he was a wonderful leader for all People little or big white or black it made no different with him.

I wrote him some time ago telling him that I had lost my job under the change of democrate to Rehublican with Mr Stranton and was buying my home and that my husband is not able to work never again and I am poor and I am colored and he took time to answer my letter showing me reinsurance [reassurance] that he would [not] let me loose my home every good President that we get Some one hafter take his life but one day there will be a nother good President and no man will be able to tuch him and that the good Lord.

I do believe Mr Johnson will make a nother good President If all the people stick with him me for one going out there and work on the pole from the time they open until they close if the good Lord is willing.

Mrs Kinnedy Please Send me some thing Personal of the President a pair of his socks or a Shirt Some thing I would like to have one of his Rocking Chairs if you not going to keep them I want some thing of his to keep in my living room for ever

PS. I always adore you for the love and care that you give to your Children they have a Sweet mother I wish you would visit me in Harrisburg I am colored and poor but Clean you are welcome any time

Please answer me back. So long for now

from

Grace Pinkney

For many World War II veterans, John F. Kennedy's assassination had special resonance. They recognized formative elements of their own life experience in that of President Kennedy. He belonged, of course, to their generation, served his country during wartime, and was decorated for heroism. In August 1943, the PT boat he commanded was rammed in the Solomon Islands by a Japanese destroyer, killing two members of his crew and leaving the rest clinging to the hull of the ship as fire burned on the sea around them. Kennedy's tenacious effort to rescue his own men became the subject of a John Hersey profile in The New Yorker, later excerpted by Reader's Digest, which helped JFK considerably in his early run for Congress. Veterans identified not only with Kennedy's military service, they also admired his lack of pretension as a naval officer, his courage and commitment to the men he commanded. Several veterans commented in their letters on the terrible irony they saw in JFK's death. He had survived a brutal theater of war, they observed, only to die on the streets of an American city as President.

TELEGRAM
WD064 PD
NEW YORK NY 22 323P EST

MRS JOHN F KENNEDY
THE WHITE HOUSE
I REALIZE THE UTTER FUTILITY OF WORDS AT SUCH A TIME, BUT
THE WORLD OF CIVILIZATION SHARES THE POIGNANCY OF THIS
MONUMENTAL TRAGEDY. AS A FORMER COMRADE IN ARMS, HIS
DEATH KILLS SOMETHING WITHIN ME
GENERAL MACARTHUR.

EDITOR'S NOTE: During the war years, Kennedy had no love for General Douglas MacArthur, who commanded Allied forces in the Pacific. MacArthur telegrammed Mrs. Kennedy less than two hours after JFK's assassination.

NOV. 27, 1963
WALTHAM, MA.

My Dear Mrs. Kennedy,

I know you do not know me personally but I am one of the crew members of your Husbands P.T. 109

I wanted to send a telegram with my condolences but it seemed to cold & short.

Your husband was one of the greatest men I will ever have the honor to meet. I am only sorry I never had the honor to meet you, his wonderful wife.

I know the way he went was horrible but the Good Lord has mysterious ways of calling us to him.

You and your children are lovely & the whole world's heart went out to You & them in the greatest trial I hope ever shadow's your door.

I hope my feeling's are legible in the above lines, but it has always been hard for me to put my feeling in writing & right now my heart is a little broken.

In my mind & heart you were married to one of the greatest men in our time & time to come. I do not think he would have hurt a soul intentionally he was that much of a Christian & if the rest of the world could have been the same there would be no more tragedy's.

I want to wish You and your children all the happiness in the world, and thank you for taking the time to read my humble message.

Sincerely,
& Love to the Children
Charles A. Harris

Charles Harris (far right) with John F. Kennedy at Brandeis University in January 1960. Eleanor Roosevelt is in the background. Kennedy announced his candidacy for the Presidency earlier that day.

PORTLAND, OREGON
NOVEMBER 25, 1963

Dear Mrs. Kennedy:

I feel the country was lucky to have had your husband at all. It might never have been. I too, was brushed by death many years ago in the South Pacific War. Ever since, I have felt the years to be bonus years . . . good years that might never have been. Years that have produced a fine wife and three lovely children.

I suspect the late President, a combat veteran, might have had moments of like reflection when you chanced by or as he watched the children at play.

If that destroyer of long ago had veered a fraction from its ultimate course, the world would have never had John Fitzgerald Kennedy. Maybe the years since that night have been his bonus years . . . his years of grace.

I know they certainly have been for the peoples of the world.

Sincerely yours,

Austin R. Matteson and Family

....❋

DRAPER E. WHITT
RICHMOND, VA

To A Dear Beloved First Lady:

It was with grief anger and disbelief that I heard of the death of our beloved President. You have my very deepest Sympathy.

I have a letter from the president, I will cherish it as long as I live. When the president received the nomination for president I got a badge from Senator Yarborough's office.

I am a Methodist and worked in Cradle Democrat. When I visited my home in Chatham County North Carolina Congressman Harold D. Cooley's district Friends of mine republican and Democrat, asked me was I going to Vote for a Catholic. I said no I was Voting for a man I felt to be best qualified for president.

Then I thought back to the 1928 election between Hoover and AL Smith. My late Granfather pin Al Smith badge on my left breast I have long realized the injustice done to Gov. Smith.

So on my return to my home here in Richmond, Va. I wrote editorial to my paper. In Siler City, N.C. Express my opinion on the political issue. I asked this one specific question.

Did any one question the loyalty of a Catholitic during war as we served together. Shed our blood together. Shared hospital rooms together. That I felt one should not be denied to right to be president if he was cabible. Regardless of his religion.

I had many friends who were Catholic. When we were preparing to sail for France on the invasion of Normandy they went with me to Holy Communion service. I with them to mass.

I to was shot as was the president. In the head by a sniper in left temple bullet coming out back of my neck. I was temporally blind and partially parlized. I am much better But total disabled.

I don't think we have ever had a president as courageous and has tried so hard to help the down hearted and depressed people.

I feel he has all in his power to full fill his campaign pledges to the people. Not only has America lost a great leader and Champion of the people. But the entire World has lost a great leader.

I am sure he will go down in history as one of the greatest leaders in our time. But above all he has done I have been more impressed by his attending Church on Sunday regardless of where he was.

As I have told my wife and two sons I believe he is the most devoted Christian presidents we have ever had. The president and I were about the same age each being born in 1917.

It is a sad day in our nation. There is little one can say. Except you have my sympathy. I know having lost most of my people Time alone can heal a broken heart.

You and yours have our very deepest and heart felt sympathy. May God in his wisdom and divine love comfort you and yours. Give you strength and courage to bear this horrible burden. Some day in another world we will all be reunited with our master Catholic, Jew Protesant and all other. Colored and white.

> May God bless and keep you all
> Draper, Lucille, Nelson & Steve Whitt

INGLEWOOD CALIF

Mrs. John F. Kennedy:

I have never written to a public figure in my life, but over the events that have happened I feel compelled to write to you. I know that what I can say can have very little recompense for you but I thought you might like to know the feelings of one person among the many who love our Country. I joined the Marines in August 1942 and spent 31 months in the Pacific with the First Marine Division. I never met your husband but I have always felt a deep bond of friendship and brotherhood toward him because we served in the same theater of war. I have never had the good fortune to visit our Nation's Capital but I have always been proud of the reminders there our past history and great heritage. No one could have been prouder than I when your husband was elected President. I said at last in my generation I have seen a man elected who will stand up and say we are Americans and we are here to stay. I watched his Inagural Address and when he said: "The Torch has Been Passed" I had tears in my eyes. I have had tears in my eyes for the past five days and hope some of them can in part pay for your own. My sorrow is deep for I spent most of my life in the South and in Chicago. I have seen hatreds of all three kinds: Racial, Religious, and Color. Since your husbands funereal I have heard people here talk against him because he was Cathalic. I stayed home Monday and watched the sad ordeal you had to go through. Your courage and dignity will never be forgotten by me. I hope you will pray for the same dignity to be bestowed on the rest of our nation. This nation of ours was founded by people of Europe who were seeking freedom of Religion and the right to Worship God in the manner of their choice. Even though this great move- ment took place three hundred years ago and even though our country has bled through many wars to achieve and protect this Right and Liberty we are still in the midst of prejudice and bigotry and misunderstanding. I am only an average American as far as education is concerned but I am an American who believes in the dignity of man. I was born in Tennessee and have lived in Kentucky, Illinois, and California. I put my life on the line to fight for the American way of life and at the age of 48 I am ready to do it again. Your husband did and I could do no less and call myself a man. I am proud to have been privileged by God's Grace to have lived in the same generation as you and your husband. What a precious heritage your children have to look forward to just to walk in your footsteps. I would

like to borrow a line from Christine Rossetti and add a few of my own in
memory of a friend and a warrior.

"Sing no sad songs for me
But sing one for us all
If hate and prejudice
Against our Fellow Man
Is Now beyond Recall."

In Loving Memory
Harry T. Emery

.... ❧

Dear Mrs. Kennedy:

I hardly know how to start this letter as I know you have received
thousands of letters from people all over the World offering sympathy
and condolence to you over the loss of your Husband & my President &
the friend to all the World.

One incident is sharp in my mind. I have 4 girls and one boy; he is
nine years old. The Girls are all Older.

About a week ago I went into my son's bedroom after the Family
had retired for the day. my son had Pres. Kennedy's inauguration speech
record on the record player and a picture of Pres Kennedy on the Pillow
beside him and a picture of "John" John crawling through the door in his
fathers Desk.

There were tears flowing from his eyes and he was sobbing as if his
heart would break[.] I asked Purwin what was wrong. His answer was
what will Pres. Kennedy's Family do. His answer did something to me.
To think that a nine year old boy's first thought would be of others in a
crises like this. It made me feel extra proud of him.

I feel a little closer to Pres. Kennedy than many perhaps.

I was in the South pacific at the same time he was. I was on a torpe-
doed LST ... we were sunk one day after his Torpedo boat was rammed

only 100 miles away I was on Rendova right next to his base and I fought through all the Solomon islands from Guadalcanal to Rabaul on through New Guinea and the Phillipines and was Wounded above the city of Manila. Although I was in the army I would have been proud to have Pres. Kennedy as my commanding officer. I know I have said very little to help or encourage you. But I know this: you are one of the most courageous persons I have ever known and a fitting memorial of Pres. Kennedy one of the greatest men this World has ever known

God bless you and your family

> Donald W. Sanders
> Mattoon Ill,
> Democrat
> Prec. Commit.

N.Y., N.Y.
[JULY 30, 1964, POSTMARK]

Welcome to you, Mrs. Kennedy I hope you like your New York home. The City at present has a rather bad reputation. It has earned it, I suppose. But there's a lot of good here, too. In time things will even out.

I'm an old infantry ex-tst Sergeant from the same war your husband fought. I cried too when the thing happened. No soldier should go that way. It's disgraceful. A soldier also has done his duty and fought a war honorably, and lived through it still, has the right above all others to collect his Social Security and live out his life in peace. For if he _wasn't_ there _who_ would collect their Social Security?

I know how you must have loved him (I lost someone that I loved very much, too) He was a giant. A man like Saul—head and shoulders above the crowd. I saw you on television, how you looked at him. And he earned it, too.

It's an awful spot for you. So young and so beautiful, you are. But where will you ever find another John Kennedy? It is such a terrible waste of natural resources, your youth and your beauty.

Your children are adorable. That photograph of young John John saluting his father should become a part of our history. Or does it hurt too much? Caroline. Anybody who doesn't love Caroline is a nut.

I live over on the East River , but sometimes I walk over to Central Park. And I'm an agnostic. But when I pass your house I will say a prayer to your god that you may find some happiness, or at least some peace of mind.

Bless you, *Mrs President

Grady McNeil

*"Madame" president I guess is the proper address. Never had too many presidents in my family. Sort of rusty about it.

---

*L*ess than twenty-five years separated World War II and the Kennedy assassination. For many letter writers, losses sustained during the war hung in the backdrop of the condolence messages they sent to Mrs. Kennedy. Some who had witnessed atrocities found the pain of those experiences revisited upon them. This was especially true of Holocaust survivors, many of whom made reference to their experience in consoling the President's widow. Others recalled the anguish of losing their sons, brothers, and fathers in battle. Grief persisted, they reminded the President's widow, and continued to challenge them many years later. Their sorrow was not only revisited but augmented by the violent death of the President.

COMMERCE, TEXAS
DECEMBER 1963

Mrs. Jacqueline Kennedy and children,
Dear Mrs. Kennedy,

With others, we want to express our words of deep sympathy, with you and yours, In the Loss of your Loved one, the President of our United States. We feel we to have shared part of your great loss, in the loss of our only Son Staff Sgt Robert Wayne Hammonds we lost in World War Two, on Anzio Beach head in Italy. We had his body brought home for reburial, and I relived it again with you, when they tenderly handed you The Flag. only those that have suffered this sorrow can feel the emptiness and Loneliness it leaves. But our Dear Son will never leave us alone if we trust Him, and Dear, you and your Loved one, knew so much of Happiness, and Had your cup of Joy, and Sweet Memories are one <u>Great</u> <u>Gift</u> <u>of</u> <u>God</u>, that death can not Destroy.

May God Bless you and Your Family.

From Gold Star Parents.

Mr. and Mrs. R. T. Hammonds.

FAIR LAWN, N.J.
NOVEMBER 29, 1963

Dear Mrs. Kennedy,

This letter is to express our heartfelt sympathy at your and our country's tragic loss. Your husband was a great President and a great leader and his loss will be felt by everybody. May the fact that he arose the nation and the fact that his policies may be carried out be a small consolation to you.

As one who has lived thorough a Nazi concentration camp and saw deaths daily from hanging, shooting, starvation, yet at no time did I feel such grief as at the passing away of our great President.

May the Allmighty bless you and protect you and your children and keep sorrow away from your door. May the soul of your beloved husband, father of your children and our great leader, rest in peace.

With deep Sympathy

Morris Starr

VINELAND, N.J.
1/21/64

Dear Mrs. J. F. Kennedy

With the best regard to you, and your loved ones permit me to express our depest Sympathy in your berivement, as we that is Me and my Vife have goon thrugh it 10 years ago, so we know how much you went thrugh, but much more. We lost our Son June the 2nd 1944 he was 21 years of age he was a Pilot in U.S. Air force, and also lost 2 Brothers in Europe fighting the Nazist, so that is why we know how big is your <u>sorrow</u>, we are just plain people living on our Social Security, when that tragic day came, for 4 days we could not control are tears, and again January 19th we saw the Celebrated Mass on Television. I could not stoped the tears, because we <u>Love</u> your Family so much, and you Mrs Kennedy you done so much for the White House in restoring the Originals in such a short time, our only wish is that we will be able to visit the Arlington Cemetery this sumer and pray there, forgive me for my English, I had no schooling in this Country I was 22 years of age when I came here 1913.

With Love to you and your children

We Remain Respectfully yours,
Adolph and Mary Macho Sr.
Vineland, N.J.

CLEVELAND HEIGHTS, OHIO
JANUARY 16, 1964

My Dear Mrs Kennedy and children,

There are no words available to us in our great English language, to express to you and your family, even after some time has lapsed, how we feel of your great loss. All we can offer to you is prayer, and may the bleeding heart of Christ give you strength to carry your cross.

You see, we too, are trying to carry a cross. On Memorial Day of 1963, our beloved nephew the only son and grandson of 5 children, accidentally took his own life. Many correlaries are manifested in his death and that of our Dear President Kennedy. The weapon—a gun. A gun which was a toy—a war souvenir of Okinawa. Rusted, and with no working trigger, this child, (11 yrs old) obtained a bullet, and with determination would make this gun fire. He was sighting the gun—looking through the barrel, totally fearless, and kept striking the hammer of the gun. It is said to be a 1000 to 1 chance, but it was a greater hand than his, that proved the fact.

The suddenness of death, the fearlessness, and determination can be correlaries. You wonder where you get the strength to carry on. But you do go on, perhaps because of their prayers.

Our beloved president will be remembered the world over until eternity for the man that he was. Oh, How proud you must be! And rightfully so. And we, your people, are so proud of you.

I am enclosing a picture of our Dear President Kennedy. Perhaps you have received many, many of them. This picture was presented to us at Mass, last Sunday, January 12th , at our church of the Gesu. President Kennedy, raised the Catholic religion as the Supreme Religion.

May that ray of sunlight which presented itself upon his grave the day of his funeral, shine forever, and grant you and your family peace and strength.

God Bless you.

Fondly,
The Dominic A. Tomaro Family

Dear Mrs. Kennedy:

On the day of President Kennedy's tragic and unnecessary death, my small son and I were at my mother's as usual for Friday night dinner. I was reading a book when my step-father called and told us the shocking news. Neither my mother or I said anything, we just cried. Later when my husband came home he didn't say anything, but you could see how terribly upset he was. He just looked at us and said he wasn't hungry. We didn't eat much that night or for the next few awful days.

After supper we went grocery shopping, with my mother as usual baby sitting for us. The few people in the store did not speak, but when your eyes met theirs you knew how they felt and what they were feeling for you too felt the same way.

After we had arrived home and John was in bed and the food put away, we turned on out television set and sat there and watched, but we still did not speak.

I then looked away from the set and the wall of our den caught my eye, no not the wall, but on an 12 ½ X 15 ½ piece of paper framed with a black rim. As I read this to myself I cried silently. It reads:

> In grateful memory of Private Julius B. Adler A. S. No. 34570089 who died in the Service of his country in the Southwest Pacific Area August 22, 1943.
>
> He stands in the unbroken line of patriots who have dared to die that freedom might live, and grow, and increase its Blessings. Freedom lives, and through it, he lives—in a way that humbles the undertakings of most men.
>
> Franklin D. Roosevelt

My father had given his life for his country as did President Kennedy. As I read it over and over I also said the name of President Kennedy in place of that of my dad's. How well he fits these words.

You have a flag with fifty stars on it; I have one with forty-eight.

I never knew my dad but I am proud of him; just as proud as you are of our late President.

My husband and I are 24 and President Kennedy meant something special to us. He was young like we are, and he understood us. He was someone we looked up to and admired, an idol, someone to pattern ourselves after in his love of country and fellow man. We will truly miss him as will the rest of the people here and abroad.

Please excuse the length of this letter but it is a rather difficult one for me to write as I have a hard time expressing my feelings, especially when it comes to my father.

I guess all I am trying to say is how awful I feel about what has happened and how my father and our late President had just a few things in common: They were loved, husbands, fathers, sons, and above all, they gave their most precious belonging for their country: their lives.

May God be with you and your family.

Sincerely,
Mrs. Donald S. Cohan

<hr />

MENLO PARK, CALIFORNIA
NOVEMBER 30, 1963

Dear Mrs. Kennedy:

Like millions of our fellow citizens my family and I wish to express our deep sorrow and grief which you must feel at this time. I first wanted to wait a little longer in order to avoid an over-emotional outpour. You have shown so excellently how one can control ones emotion, therefore I shall try the same.

First of all, Mrs. Kennedy, let me assure you that all four of us share your grief, your pain, and your sorrow. We feel the late President has been very close to us, and our distress could not have been greater than if a member [of] our own family would have gone. Of course we cannot compare our grief to yours because grief cannot be compared.

My wife Pauline and I are foreign born. Both our children were born in California. Katherine is almost 17 and Douglas is 9. Pauline and I came from Austria some 23 years ago.

Just before Austria lost her independence to Mr. Hitler, I was taken by the invading Nazis to Dachau and later to Buchenwald. I spent two years there in misery. Released in February, 1940 from Buchenwald, I went to Sweden and early in 1941 arrived finally in the United States. Almost immediately I volunteered for service in the Army but was rejected because of lack of citizenship. Soon thereafter the law was changed and I became subject to the draft. Trained to become an interrogater of P. O. W.s I spent much of my overseas time in Military Government work with Infantery Division. After 22 months overseas and after experiencing the end of the war in Europe, I returned in 1945 to San Francisco. During my time overseas I had the rare opportunity to revisit the two places of my confinment and even could participate in the liberation of Buchenwald. I even had the great joy of finding there some of my friends alive and well.

The reason for telling you all this, Mrs. Kennedy, is to prove to you that death is not new to me, that suffering is an old acquaintance. I have seen many a good friend die, and many of them tortured and abused. I myself have suffered from the cruel hands of men and I like to think of myself as an unemotional man. Yet once, and only once did I ever weep for a dead friend, and that was in 1940 when I learned that one of my friends, by the name of Rudi, had been murdered by the Nazis.

I wept, not because he was my friend; I saw many a friend die. I did not weep because he was of my political or religious belief: he was not. I did not weep because he was young (27 years): I saw many die much younger than he. I wept because I knew he was irreplaceable. I knew he was needed. I scand in my mind all those I left behind and I could not see a single one as strong, as couragous, as wise and as good as he was, who could help those who needed him. That is why I wept in June, 1940 in Vienna when the Nazis murdered Rudi, the best of them all, the only one who could, at that time, lead those desperate, persecuted and troubled men.

That, Mrs. Kennedy, is precisely the reason why I am weeping now. Again a man has been taken away from us who was good, wise, noble, and kind; one who has been our hope in a desperate situation, who has shown us the way; a man who has been harrassed and yet could smile and find good things to say for his adversaries.

Words are cheap, Mrs. Kennedy, but how else can we tell you how sorry we are, how great our pain is? Only words—for the moment at least—can express our sympathy for you and your children and how sorry we are for our own sake. Yes, for our own sake. Where in all this great and mighty country is the one who could replace this man? A great leader, the greatest of them all, has left us. One who, only once in a hundred years, seems to appear among us. He gave us strength and confidence and again a trust in the future. He showed the way and the world will never be quite the same. He left his indelible mark on all of us.

You are a great lady, Mrs. Kennedy, and you put us all to shame when you showed us what courage means. But after all you have been so fortunate to have had this splendid man ten years at your side. We only had him three. All we asked was for five more years; but Providence did not grant it to us. And yet we consider ourselves fortunate that this "good man, this honest man, this kindly man" was our contemporary. He touched us only such a short time but we have already changed.

Please, Mrs. Kennedy, do include us in your grief and your sorrow. We have lost the greatest man among us as you have lost the finest husband and your children the best father.

With our kindest regards to you and your children we wish that you may have strength and hope for the future. May God be with you and yours.

Pauline Ernst

Katherine Ernst

William Ernst

Douglas Ernst

DETROIT, MICH.
APRIL 29–1964.

Dear Mrs. Jacqueline Kennedy,

I am one of those persons who lost my entire family in a very cruel way during world war two, in Europe. I thought that nothing in this world could hurt me any more, but how mistaken I was! My heart still cries out for our beloved President.

Will you please accept my deepest sympathy.

Sincerely yours
Mrs. Jean Schechter

·· 🌸

Dear Mrs Kennedy,

First of all, if I may, I would like to introduce myself. My name is Donald Farquhar. I am twenty-one years old and I am a resident of Groton, Connecticut. I am employed at Electric Boat division of General Dynamics.

Please forgive me if this letter is untimely and out of place at this particular time. It is probable that I cannot express any of my feelings personally that have not already been generally expressed by the American public. Nevertheless I had to write to you to attempt to express my personal feelings about your great husband. I know thousands of personal feelings have been expressed from Americans and people throughout the world by sending letters and sympathy cards.

This letter not only represents my feelings, but I'm sure the feelings of all the younger American citizens that I know personally.

Your husband, President Kennedy, God Bless him, was in my estimation one of the greatest human beings that ever lived and the greatest presidents in the history of the United States. I have read that your husband's Presidential Inaugural Address will go down in history compareable to Mr. Lincoln's Gettysburg Address.

As far as I am concerned I will always regard you as the First Lady in the United States. You are the most beautiful and gracious First Lady that we have ever had in the White House. If you can believe me, a part of you died when your husband did, May God Bless him, he did more by living that all of our great Presidents did together by living.

If I may, I would like to tell you one last story. My father owns a used car dealership in Groton. An old Italian gentleman lives next door and barely speaks the English Language and has a cold personality, but what he said about your husband touched me. His son was killed in World War II and he still grieves over it. He came over to my fathers business and told us that he felt worse over your husband's death than he did about his own son. He meant that statement from the bottom of his heart as I do everything I have said to you.

As you know, a person will be criticized even if he were the Lord himself. Even the narrow-minded, bigoted Republicans I know respected your husband, which is really saying something. The party to which a person belongs, in this sad case, should be immaterial, because a person should be an American before he is a Democrat or a Republican.

The President also had the youth which enabled him to especially attracted the interest of the younger Americans. It is senseless for me to carry on because I would never finish this letter.

Mrs Kennedy, if you ever hear any narrow-minded person criticize your husband this letter may help you remember how American people feel about your husband. I would consider it an honor if you would take time to read this letter. I know it is impossible for you to answer every letter you receive. I would like to meet you in the future if I may.

Mrs. Kennedy, Please believe that I meant everything I have said from the bottom of my heart. Goodbye and God bless you an your beautiful family.

Yours Truly,
Donald Farquhar

Dear Mrs. Kennedy,

This letter should be on parchment for you. But I am a plain midwestern boy and perhaps you can draw more from this simple correspondence.

I had only one contact with your husband, but even if history had not focused on him as much as it did, I would have always felt that special spot in my heart for Jack and you. I suppose it is the same "hero worship" he felt in his early life.

Endless words have gone out in his favor, but perhaps the most I could write to you is my description of the way I knew him best.

I was attending the University of Illinois when Jack campaigned in 1960. I remember all the talk about his religion and his family's wealth. But I never figured religion entered the picture. And my mother said here is a man they won't "buy off."

When Jack came to Champaign, Illinois for a campaign speech, I as so many Americans realized we were for this man. Jack stirred me with what he said along with everything else he said later on. When Jack finished and drove to the airport I chased his car six blocks just to shake his hand and wish him well, and you know even though I was the last well wisher running at full speed he sincerely thanked me.

I was in the army during your husband's term and so he has been my President and my Commander in Chief.

My father was killed in action in the Battle of the Bulge during the second world war. My mother and I received a parchment from John F. Kennedy honoring his (my father's) service to his country. My family will always treasure that parchment and we will always hallow this memory of John F. Kennedy. Now this is our way of thanking his family for his service to his country. For the Lord must have surely had a twinkle in his eye when John F. Kennedy came to this earth.

There will be much said about this man and many ways of honoring him. But you are his greatest monument; You and Caroline and John Jr.

You have given us all courage, may God bless you and keep watch on you. My fondest wish is that I might be able to meet you some day.

Sincerely yours,

Leonard L. Link Jr.

....❧

The condolence mail to Jacqueline Kennedy contained thousands of letters from those who described the loss of their parents or children. Some saw themselves in the fatherless children JFK left behind. Others recognized the enormity of the burdens shouldered by family members who had to live with the particular pain imposed by the violent or premature death of a family member. Many more drew upon the closest example they could find in their own life to the sadness they felt they saw in Mrs. Kennedy, and, indeed, felt themselves again upon the occasion of the Kennedy assassination. Still others celebrated their deceased family member's affection for JFK, reminding his widow of the pleasure many took in his presidency.

4.12.63
SAN DIEGO, CALIFORNIA

DEAR MRS. KENNEDY
"WE ARE HEALED OF A SUFFERING ONLY BY EXPERIENC-ING IT TO THE FULL."
I, TOO, WITNESSED A TRAGEDY. MY LITTLE GIRL DROWNED.
"IT REQUIRES MORE COURAGE TO SUFFER THAN TO DIE."
NORBERT A. HOERL

....❧

DECEMBR 1, 1963

IN 1962 SEPTEMBER 23,

SOME MEAN MAN Killed MY DADY Too-
HERE IN DAllAS – MY DADY WAS a
SoldRER
SANDACLAUSE DIDEN GET MY LETTER

i hoPE HE Will GET MY LETTER
i WON T A BiCYCLE –

When YoU WRITE him – TELL him

MY NAME –

MonROE YOUNG JR. II

1838 NO MAS. Street
DALLAS R, Tex.

COLUMBUS, OHIO
MAY 28, 1964

Dear Mrs. Kennedy,

My name is Patricia Boling. I wanted to write to you right after the assasination, but I didn't have the courage. I am sorry I waited so long. I am a Catholic girl, thirteen years old, and I go to St. Ladislas School.

I suppose you would say I am a person that expresses what I feel in my heart. My grief and shock cannot be explained. I dearly loved President Kennedy, not because he was the president, but because he was a man that stood for everything I hoped and believed in. He was the symbol of the America I loved. I cried so hard on Friday, the 22cd, and so did a lot of other girls. We were in the for fourth or fifth grade when he was running for pres. We used to skip around the playground at noon chanting, "Kennedy, Kennedy, he's our Man, let's throw Nixon in the garbage can. It sounds silly now, but it was our way of voting. For some reason I felt safe and secure, knowing Pres. Kennedy was the head of our country, and the first few months after he died, I was scared to death something else horrible was going to happen. I finally quieted my fears, but sometimes my imagination runs away with me. You will always be my model of a true woman. What I mean is, someone who can endure what you did and still stay calm out in public. I know for a fact, if I were you, I would have cried and cried. I know I cannot nor could not find any solace in tears, but sometimes they help to ease the hurt.

My mother died of leukemia when I was six years old. I have 5 brothers and sisters, the twins were only 2 yrs. old when she died. I had the great privilege of knowing my mother a little. But I always feel some hurt when I know my brothers (Jim + Dave, now 8 yrs.) will never know the comforting arms of a real mother. I myself vaguely remember her, but everyone says she was the kindest, most gentlest woman that ever lived. It has been seven years since she died, but Daddy still gets that faraway look in his eyes whenever anyone talks about her.

I am still young, so I don't know many things. I have a question to ask, something that has been troubling me ever since Pres Kennedy's death. Why does God let everyone so good and kind be taken to heaven, when we so badly need them on earth? I have been trying to find an answer, but I can't do it myself. Please help me to understand. I know that death is only parting a short time, compared to eternity in heaven, but still I think I shall never see Mommie as she was on earth. Sometimes I feel so sorry for myself because I don't have a mother that I cry. But I soon stop, because I have been taught that everyone is radiantly happy in heaven, and we shouldn't wish them on earth again because everything is so perfect in heaven.

Right now I am thinking of my vocation. Sometimes I would like to serve God by being a Sister. Other times I want to marry and have children, to teach them how important the phrase For God and my Country is. I will pray for you and I wish you all the luck and happiness in the world.

Please pray for me too.

Ask God to lighten my rough and crooked path to heaven, and I will do the same for you.

I love you and

God bless and keep you and the children

My sincere love and respect
Patricia Lee Rita Mary Boling

PALESTINE, TEXAS
DECEMBER 6, 1963

My Dear Mrs Kennedy,

I am so sorry that am just now sending you this card with this letter. I am also sorry that us peoples down here in the South has (bad) a State like this place called Texas. I even hate to live in Texas now, but me being Crippled all up in my right leg and arm with Arthritis in my back. I will have to lived here the rest of my life, I was born here in Palestine, Texas January 18,-1934, my parents are dead, a white man told my father to come out to his house and get his money for some ducks. My Father went out there when he got out of his car the white man shot my Father through a window with a shot gun, this white man just murdered my Father back in 1946. then in 1951, my mother died in Parkland Hospital from an operation in Dallas, Texas. I am disable to work, but I can walk. I walk with a limp, by me being Cripple, Texas is rotten, this Place is so rotten that the State Dept. Welfare wont give me a disability, isnt that rotten, when that man murdered my Father the Police did not put him in jail they just let him go. I new that they would not do any thing to him, you see, my Father was a negro just like me, but that was 17 years ago. I just can not take a man life. I would rather be dead my self, Mrs Kennedy I my self hope you have a nice Christmas with your little boy John Jr., and your little girl Caroline. I also hope your life be as happy and joy can be, I just love to have a big picture of you, Mr. Kennedy and your two children. I would love to have a big color picture of you and your family. I was crazy about your husband Mr. John F. Kennedy. I wont ever for get him, so I will close this letter, so good by Mrs. Kennedy.

from a Friend
Clinton Hale Holman—

....❖

EL CENTRO, CALIFORNIA
JANUARY 26, 1964

Dear Mrs. Kennedy:

As I watched the Television program for four days feeling the great burden that was suddenly thrust upon you and your loved ones, I never realized that within a couple of weeks I, too, would be going through a similar experience.

I am a mother of three sons all serving the country in various fields. The youngest, Roger, chose to serve with the program that your husband initiated, and as a Peace Corps volunteer was assigned to a primitive and remote barrio in Dikalongan, Mindanao of the Philippines. Here he passed on December 9th, 1963 suffering a reaction to the malaria suppressive they were required to take.

In Roger's second to last letter written I would like to quote what he had to say about your husband upon hearing of the assassination. "The assassination of President Kennedy is a great and trying lost to our nation and to the world. He was a champion leader of Peace and freedom, guided only by honesty and respect for the individual. In one respect it is a loss, yet on the other hand we should be grateful for being the recipients of such firm leadership, though it ended abruptly. I will never forget the morning I heard the news. It was November 23rd and was one of those grey overcast mornings when one thinks it is earlier than it really is since it is darker. Peter and I had gone to Esperanza for the district track meet. I had rolled up my mat and mosquito netting off the floor and had walked over to the school to a morning breakfast of rice and fish with the teachers and children when I was told the news. I thought my Filipino friend had not gotten the news correctly, but I went back to the house where we were staying and woke Peter to tell him. Neither will I forget sitting on the dirt floor of the tailor shop listening to the radio as the rumors were confirmed and tears clouded our vision. That evening we went home to listen to the Voice of America broadcast. In order to obtain better reception we turned off our one electric light. The house was full of friends that had come by to listen with us. And the Voice of America came in strong and clear with a background of Chinese music. No one spoke during the hour-long program, as we listened to the news of the tragic ending of our President."

In Roger's frequent letters written home he would tell of the great beauty he had found in another's land and the joy in getting to know and understand another race of people. With all of this rich material, one of his older brothers, Rick and I are putting together in a book. We hope not to convey so much the idea of a memorial to Roger but rather to perpetuate the involvement of more people to the extent of Peace Corps activities. However the essence of the message will prove in the last analysis that death was not the victor.

And I would like to share this thought with you, Mrs. Kennedy, from St. John 17:4, which has brought much solace to me.

"I have glorified thee on the earth: I have finished the
work which thou gavest me to do"...

> With God's blessings and prayers for us
> both,
> Dorothy McManus
> (Mrs. Dorothy McManus)

···-❈

Dear Mrs. Kennedy,

This is not a Merry Christmas and Happy New Year message by any
means; however, I am intentionally writing at this time, as I know how
you are feeling at this time of the year.

My mother is in very low spirit lately and talking about Mr. Kennedy
doesn't help the situation; my friends are all excited about the holidays
and aren't interested in me because I sincerely wish the next two months
were passed. So I had to write to you immediately.

I have been a politician since 1952, when I was four years old. I told
all my friends to vote for Stevenson because my uncle said so. (He was
the Democratic precinct captain) Before, 1960, I was interested in a man's
party rather than his personality. That fall my mother had a huge picture
of J.F.K—the man for the sixties in our picture window. My father said
"He can't miss—he's Irish, Catholic and Democratic". At this point I
became interested in your husband not as a presidential candidate but as
a man. During November, the family watched the campaigns, the election, and in January, the inauguration.

When the news reached us at Mother McAuley, the whole student
body started to pray first that what we heard was just a rumor and, later
that God would spare our president, our leader. Ill never forget that
day—it was raining in Chicago, but I didn't even bother putting up the
umbrella which hung from my arm. When I arrived home, I watched the

T.V. broadcast and I cried. My mother came home from school in tears and joined me in front of the television. We watched until the stations stopped broadcasting for the night. The next day, we watched Mayor Daley's speech to the city council as well as other memorial speeches.

There was a certain numbness the following week; everyone was thinking of the President but none were able to speak his name.

Often, I have thought of Mr. Kennedy and cried. The tears, however, are partly for the other great man who died on May 16, 1963—my father. I have made many comparisons between them. My father was not wealthy, he was not a great statesman. He did serve on the Solomon Islands in W.W.II and came home after suffering from malaria. He was not the commander of a PT boat, he was Staff Sargeant Joseph A. Kilmurry, better know to his friends as "Irish". He longed to go back to Ireland to visit West Meath which he left over thirty years ago; your husband did make the trip. My father spent hours in a rocking chair which he said was "just like Jack's"; but he was not resting his back—he was dying of cancer. You love your husband—I love and miss my father. We have much in common this Christmas.

I want you to know I'm praying for Mr. Kennedy's soul as often as I pray for my father's which is at least once a day.

My only regret is that I never came to Washington. Now, I have no desire to see the White House or the Capitol Building but I have been promised a trip to Washington so I can visit Arlington Cemetery and say a prayer over the grave of the greatest of Presidents and one of the greatest men of the world.

I don't expect an answer, as I know you are busy but, if at all possible, I would appreciate a prayer card.

With love and respect,

Ellen Kilmurry

P. S. Enclosed is a page from our school's newspaper—the Mother McAuley Inscape

TUCSON, ARIZONA
NOV. 25 1963

Dear John Jr.

Altho I am sixty five years five months & five days older than you, we will both have one thing in common, a birthday to remind us of a tragedy that happened to us in the past.

Altho I did not vote for your father when he was running for president I have learned to admire & love him as he grew in his job, not the deep love that I know you & your sister & mother had for him, but the kind of love that is mentioned so often in the bible, where it says we should love our neighbor as our selves, those things you will learn as you grow older.

Your father was shot & killed on Nov. 22 & layed to rest on Nov. 25th which is your birthday so you will always be reminded on your birthday of what happened on your third birthday, but on the other hand you can be reminded that there was more heads of states & more heads of countries at your fathers funeral (and rightly so) than any other man ever layed to rest.

My father also was killed on Nov. 22 in a saw mill accident when I was five months old, also that same day was my mothers seventeenth birthday, so she was left a widow at seventeen years of age with a five month old son, but a little over a year later she remaried & had ten more children, & while they were always very poor, the one thing she tried to do was raise us all to live for & believe in God, & I will always thank her for that, also she out lived her second husband & she is still living & in her eighties.

I of course got married & raised two fine girls & a fine son, & our son like your father was in the second world war, & in 1944 on Nov. 25th which is your birthday, our son was killed in action, so you see I will never forget your birthday or the date your father was so cowardly killed & may God bless you & your sister & mother & give you the strength to carry on & when you grow up I hope you will be elected president & take over where your father left off & finish the job that he started.

A friend
C O Carriker

....❧

IRVINGTON, N.J.
JANUARY 11, 1964

Dear Mrs. Kennedy,

I buy all these magazines and books with pictures of our dear President, and you, and your children, and today I bought a memorial issue of Look; I felt once more that empty sadness and I cried—again. But I know, from an experience when I was twelve—I am now forty-six—that one remembers forever, but not with the pain you must feel now.

It was St. Patrick's night thirty-five years ago and my mother, who was thirty-five, and my father, born in Tipperary, Ireland, were dancing in the living room when my sister Eileen, age eight, and I, age twelve, went to bed. In the middle of that night, I awoke to look into the white stricken face of my grandmother, who entered our bedroom to get a sheet on which to place my dying mother, who had hemorrhaged her arterial blood, and then my sister and I knelt in our bed, shivering and praying, and a little later we went downstairs to the living room and watched her pass from this world. Next morning, I remember forever, my father's bloody fingerprints on our white front door, put there when he dashed outside for help. That was so many years ago and the pain and terror has faded away but the love and tribute contained in this vivid memory is fresh and clear forever. My father, who has never remarried, is seventy-four years old, and I will, I must, remember him when he is gone as a wonderful, slightly enfeebled old man, but my mother, always gay, young, and courageous to me is frozen in the time of youth and splendor, and I will remember our wonderful late President in this same, comforting way.

I carry this poem in my wallet; it is a poem on death by Emily Dickinson and I consider it the best of any advice ever offered on the way we all might think of Death, not with morbidity facing an endless or immeasurable time but as a renascent time for love.

"The Bustle in a House
The morning after Death
Is solemnest of Industries
Enacted upon Earth—
The Sweeping up the Heart
And putting Love away
We shall not want to use again
Until Eternity."

Margaret Berkery

··※

The condolence letters often reveal the writers' efforts to construct their own lasting memory of John F. Kennedy. Even as they watched the official mourning and commentary, Americans reflected on how they would choose to remember the President. Drawing upon their firsthand exposure to JFK or, in some cases, a memorable encounter with other admirers of the President, these writers shared vivid recollections of Kennedy.

INDEPENDENCE, KANSAS
JANUARY 14, 1964

Dear Mrs. Kennedy,

I realize that this letter will never be read by you, personally, but I do know that the general idea of this letter and all the others will be conveyed to you by the means of your secrataries.

I consider myself so lucky to have met your husband. Of course it came as such a blow when he was killed. I was very much hurt, as was everyone else. But when I find myself moping around I stop and remember a very cold and windy November night of 1959. Senator Kennedy came to Hays, Kansas, a small [town] in western Kansas, to speak at a dinner. The Jefferson West (St. Joseph Grade School) Band was asked to escort him to the place of the dinner. After we arrived at the school, I remember him standing there in a navy blue overcoat, and his eyes were sparkling, a grin was as wide as his face. It was such a warm smile. Oh, how lucky you were to see it so often. He smiled all the time that our little band stumbled through the Naval Hymn. The band sounded petty bad, but he told us that we had done a fine job. This made us all feel as if we were the best band in the United States. He shook each of our hands and chatted with each of us. We each received a candy bar from him for our "serenade." I ran to the end of the line three times so I could shake his hand just once more, but Sister soon stopped that. This is the way that I like to remember him. I'm sure he wants it this way. I just thank God that I got the opportunity to [meet] Mr. Kennedy.

I want you to know that you and your children are remembered everyday in my Masses and rosaries.

Sincerely,
Susie Anderson
St. Andrew
CYO Member

MR. IRVING SILVERSTEIN
ROSLYN HEIGHTS, L.I., N.Y.
THANKSGIVING DAY, 1963

He stood in the open car,
Mrs. Kennedy,
the day was raw and cold. It was November. A chilling rain made the waiting for him difficult for Susan age 9 and Ellen age 6. The home-made

placards were softened and moisture had taken its toll of the lettering "We want Jack." Yet we waited.

He stood in the open car, hatless, the momentary cheer, he turned toward us, I shouted, "DON'T CATCH COLD!," he smiled, reassuringly, caught my eye for a flicker of a second and he was gone into the night.

I have my remembrance of him.

We as a family on this day offer you our prayers, our love and our wishes for your future happiness.

Irving, Sophia, Susan, Ellen

HAVERFORD, PA.
JAN 14, 1964

My Dear Mrs. Kennedy,

I knew your husband while we were at CHOATE School together.

He and his brother, Joe, were in a room in, if I recall, the 'East' Cottage, a few doors from the single room I had. It was during the 1932 Presidential campaign and I often recall that Jack and Joe seemed to be the only 'boys' at Choate that knew F.D.R. would be elected! Both boys were very popular with everyone. I recall rough houses and good fun originating from them all the time.

One incident which I recall of Jack's wit was when we had all ordered ice cream on many a night and when they arrived your husband said "they must all be yours Van"—for the tops of the cartons were either marked 'Van' for vanilla or 'Choc' for chocolate!

I wrote to him during the 1960 campaign recalling our Choate association of over 30 years ago. And I received a reply from him mentioning this. I also wrote to him after he became the President to include him in our prayers and to express our confidence in him in his great responsibilities. Ralph Dungan replied to that letter.

I literally worshipped the ground he walked on and from the time he began to campaign for the Presidency I read every word he said or wrote and got home in time to watch his T.V. press conferences.

The White House, our Country and the World will never be the same again. He was truly a great and dedicated man.

My most sincere sympathy to you and your family. I know that the many happy memories along with your religious Faith will give you the continuing Strength to carry on in these trying times.

I have the honor to remain,

Respectfully and Sincerely,

Ed Van Dyke

HOLLYWOOD FLA
FEB 14 1964

Dear Mrs Kennedy

Please accept my sympathy for the great loss you have suffered.

Having been co-pilot on the Caroline all thru the hectic days of the primaries, I got to know President Kennedy as very few people ever had the privilege to know him, and grew to respect and admire him to a point of awe.

The nine months that I spent around him helping in my own little way were the greatest months of my life and highly treasured in my memories.

Our prayers, my family's and mine will continue along with masses, as we know of no other way to show our respect at this time.

May God bless you and your children and keep you well.

Sincerely,

Roland Dumais

EDITOR'S NOTE: *Kennedy used a private plane,* The Caroline, *while campaigning during the election of 1960.*

IRA SEILER, M.D.
SPRINGFIELD, VIRGINIA
NOV. 28, 1963

Dear Mrs. Kennedy,

Words alone can not express the sorrow I felt on learning of your husbands untimely death. Even as I write this letter I can not fully except it as fact. In my mind he lives on and will continue to live on as a symbol of a man of peace.

Today, on Thanksgiving, I keenly sense his death for it was just three years ago today that I forced my breath into the lungs of his newly born son. I met your husband only once after this but the part I played in saving his son's life gave me a feeling of deep closeness to your husband. I only wish I had been able to give my life in place of that of your husband. He had so much to offer.

Your grief is also my grief; your loss is also the world's loss.

Sincerely,

Ira Seiler

ROSEVILLE, MICH
JAN. 18/64

Dear Mrs. Kennedy,

I've waited and thought about writing you for such a long time at least it seems an interminable length of time since that horrifying day. I do not write to you to deepen your hurt but perhaps to lessen mine.

I met your husband when he was campaigning for the Presidency and I shall never never forget the dynamic personality of his. He spoke in the pouring rain and people, drenched, and cold begged him to keep talking. One elderly gentleman behind me said so softly, "There goes our next President, God bless you my son." I saw the tears in his eyes as they are in mine as I write this.

Our son is 12 and I shall remember the hurt in his eyes as each day it was his task to put the flag at half mast for his school. Each day he was reminded. He was quiet about it all but one night as I went upstairs to tuck him in there was our flag, folded as they folded it on that fateful day, placed over our Bible. Each day his thoughts were with him. He never asked us why? I suppose children have such a greater faith in God that they do not question Him as we sometimes do. He (my son) has collected every picture and momento he can find. Our family attended the Memorial Masses and we felt as if we lost a member of our own family.

I know that what I say has been said a million times over and cannot change what has happened. I also know God had a reason and time only will expose to us what that reason was.

We visited Washington in 1962 and when we saw the White House we knew why he had worked so hard to get there. I never missed a television press conference or an appearance of you or your husband. I stayed up all night election night as your family must have done. I was so happy when he achieved his goal. If only God had given him a little more time.

I thought writing this would help take away some of the ache in my heart but it is still there. I suppose time will lessen mine and the people of America's sorrow but I shall never forget him.

I am a young mother, your age with 2 children also and can imagine how long each day must be for you as I find it impossible to imagine my life without my husband. You are a very courageous woman. You showed great restraint and dignity in your sorrow as you knew the whole world was watching. I shall keep you and your children, as I will keep your husband in my prayers always. I wish you to have a happy life in the future. I do not mean now but you deserve to have happiness once again.

This letter comes from the hearts of our entire family from my husband Tom, my son Glen, my little girl Nancy and myself Doris. May God bestow His blessings upon you and your lovely children.

Sincerely
Mrs. Thomas Poberezny

Dear Mrs. Kennedy,

I wish to Extend to you and your family my deepest sympathies on your loss. Most Americans this past week have felt as if someone from their own family had gone. To us John F. Kennedy was an older brother, Strong, honest and good. To me particularly, John F. Kennedy was a friend. Although we never met, I felt as though I had known him all my life. I was given by fate the ability to impersonate his voice & to copy his gestures. I sincerely hope that a part of what I did found its way to him and gave him and his family a few pleasant moments.

I guess as an adult I know too many words & therefore have trouble picking out the right ones. The News papers reported that when John Jr. was told of his father's death he said "Now I have noone to play with" Sometimes children say it right—We have lost our big brother. We must now make him proud of us.

Trusting that you and your family now receive nothing but health and happiness.

> I remain
> Respectfully Yours
> Vaughn Meader

EDITOR'S NOTE: *Vaughn Meader was a comedian whose ability to impersonate John F. Kennedy led to a starring role in a 1962 comedy album,* The First Family, *lampooning the President and the large extended Kennedy family. It sold four million copies in four weeks, and at some 7.5 million copies became one of the bestselling albums of its time. Asked at a press conference whether the parody produced "annoyment or enjoyment," Kennedy laughed and said Meader "sounded more like Teddy—so he's annoyed." The record was pulled after President Kennedy was assassinated.*

Dear Mrs. Kennedy,

About ten years ago, when my dear mother died, we received all kinds of lovely notes telling us things about my mother and about a side of her personality we didn't know. I have always treasured those little notes of love.

How much you must have gathered in the monumental bags of mail the whole world sent to you out of love for John Fitzgerald Kennedy!

I was in the hospital at the time recovering from a coronary thrombosis. Every nurse as well as the patients wept for you and for us.

But what I decided then was to tell you of an experience we had while in Ireland in 1960.

We had left our hotel after breakfast and were trying to get our car started. It being a rented car, my husband was not familiar with its idiosyncrasies. It just whirred and whirred and would not start. This was in the city of Cork.

I noticed a man with a very wide broom, who was sweeping the sidewalk. He glanced around when the car made so much futile noise and seemed to want to come over to speak to us, but hesitated and then went on with his sweeping.

After much more of the same aggravation, the sweeper leaned his broom against the building and walked over to us.

"Are ye havin' a bit of throuble?" he asked. Then he asked me if I had a bit of a rag. I fished around and gave him my cleaning rag for my eyeglasses. "Put up the bonnet." he said. After cleaning the sparks he smiled delightfully and bowed. I said, "Look here, let us do something for you, too." But he would have none of it. Instead he asked this question, "D'ye think John Kennedy will become the president of the United States?" I replied, "Of course." Then, with utter charm, he leaned his head over side-ways in my direction and said, "An' do you think, he has a <u>chance</u>?" I laughed merrily and said, "He has the best chance in the world in our country." Then we insisted on his taking something and going to the nearest pub to have a toast on John Kennedy.

You would have laughed to see how quickly he parked his broom and swayed jauntily down the street . . . no boss to say him nay . . . and I doubt whether he'd have listened.

When I go back to Ireland this summer, God willing, I will look him up and buy him another drink.

So God Bless You, Mrs. Kennedy. We love you dearly.

Another Irishman,

Peg Chapman

....✤

Among those who took the time to write condolence letters were many who sought to provide emotional support to Mrs. Kennedy. Their messages contained advice, poems, quotations from the Bible, reports of remembrances of the President, and inspiring passages from works of literature. A great many letter writers sought to provide more practical assistance, including an "old soldier down in the West Virginia hills" who offered to let John Jr. "run wild for awhile or learn to turkey hunt."

11/22/63

Dear Mrs. Kennedy

I'm nobody special, just a student who, I'm afraid hasn't seen a great deal of life.

But I've been told that if someone is in love and has been in love, he will have enough memories to last the rest of his life. It might be true.

Alden Diaz

....✤

My Dear Mrs. Kennedy:

In times of stress God deems it wise to send a man of greatness and courage to lead our Nation. Washington, Jefferson Lincoln, F. Roosevelt and John F. Kennedy. I wanted to write your beloved husband and tell him how much he reminded me of President Lincoln and that one of his greatest achievements during his term of office would be the passage of the civil rights bill. I prayed the Rosary for you constantly during your great loss, because I know your sorrow was as heavy as Our Blessed Lady felt when our God was put to Death. The following lines are from the writings of Helen Keller and were a consolation to me in attempting to understand why.

> "Let us not weep for those have gone away when their lives were at full bloom and beauty. Who are we that we should wish them back? Life at its every stage is good, but who shall say whether those who die in the splendor of their prime are not fortunate to have known no abatement no dulling of the flame by ash, no slow fading of Life's perfect flower."

The Civil Rights Bill will be passed and History will regard President Kennedy as a man of great courage, a champion of the people and a strong advocate for peace and brotherly love.

Your courage during your great sorrow was truly magnificent and I pray that Our Blessed Lady will always walk with you as She did at that time.

God Bless You.

Sincerely yours
Beatrice Joan Bond
(Mrs. Wm. Leslie Bond)

This montn just one year ago I Lillie Mae Lewis made and sent Little Carolyne a hand made DOLL for her 5th. Birthbirth date On July 22, Of this year I recivird an answer from the (our greastest Pres. of all times thanking me for his daughter gift along with a card of Which he himself autographed also a picture of mrs Kennedy and the two childr

To_day with a heart filled with exceedingly sorrow I can;t fine words suitable to really express to you and your children Mrs Kennendy as well as his immediate family. Truly hope that you realized that ther is a God that ruled the earht and heaven promised to put not upon us no more than we could bear.

Yes I am a negro woman and to I am christian an we all are human. So we much feel each other cares and woes. Your husband and our Late Pres. in my heart seam like twin brother as I am only from May 17. to May 29. older than he was.

I would like so very much to be Little Carolyn Negro Godmother .If only had the privelege to remember her on all holiday and as she grow up she too could only remember by my cards if nothing more I shall always keep your picture.

May God forever bless you and your family Mrs Kennedy

Respectfully Yours
Lillie M Lewis

••❈

Dear Mrs. Kennedy:

Through the days, the weeks, the months, my heart has gone out to you in the loss of your beloved husband and our beloved President.

I too lost my loved one in a tragedy, not comparable, of course, but an experience so calamitous that even the many years have not dimmed its memory.

I am a survivor of the TITANIC disaster in which my husband, Henry B. Harris, gave up his life.

I have wanted so much to show you in what high esteem I hold you and know no better way than to send you one of my most precious possessions.

It is a tiny beaded purse with a gold top that was given to me by Lillian Langtry when in 1901 my husband managed her American tour. She claimed it had belonged to Marie Antoinette, but in any event it is a delightful little thing and I'd like you to have it.

I won't presume to send it to you until I know you will accept it, so please let me know where it may be sent so that you will personally receive it.

In highest regard,

Irene Harris

Please forgive type. My writing is not too legible—
I. H.

Dear Jackey & children

I Just don't know how to put it in words what a wonderful and coragious person you are. I also loved the President a great deal. I shook his hand when he was in New Castle Pa. It made me feal so good and I am 62 yrs old no wonger the younger folks loved his so much. You must excuse the writing and spelling as I only had 6 yrs of school. Then My Father died so us children had to find work. I am a granmother of 15 grandchildren I had 6 children 3 boys and 3 girls.

So I just wanted to do this picture in cross stitch of your late Husband I just didn't have a frame for it so would you please except it as it is with all my love you can keep it and put it in the Library. I am also making one for each of my Sons as they sure loved Jack and worked hard for his election. I hope and pray you can read this. I did want one of my Sons to write for me.

But they said it wouldnt be rite so I should do the best I can.

Hoping you will enjoy your new home in N.Y. May God Bless you and your family I remain just another Friend

Mr & Mrs Andrew Vrabel

BATESVILLE, INDIANA

Dear Mrs Kennedy & children

this letter may have started plain but my family & I think of you as someone good plain & honest like ourselfs. We miss our wonderful President so verry much. I wonted to tell you that my Mother & Dad are old & sick. But Christmas day there was a small Gift wraped pacage for me under the tree. when I opened it, it was a plate with your husbands picture painted on it. Mother had crocheted with her Crippled fingers a little frame & ribbon all around to hang it by. it's a wonderful Christmas Gift. One I will Cherish all my life. good luck to you & your children & God bless you Always.

Mrs Jerome Paul

TRENTON, NEW JERSEY
JANUARY 16, 1964

Dear Mrs. Kennedy,

I am only thirteen and I know you are well educated, but I still feel I could give you some advise.

I have been operated on four times for polio and I am now recouporating from a broken hip, but I know you too have problems so I will tell you my remedy for smiling and happiness. Alway sing "You Gotta Have Heart" from <u>Damn Yankee's</u> and I think you'll be happy. I doubt weather or not you'll read it, but the aid whose reading this letter; it goes for you to. But could you at least tell Mrs. Kennedy this? Thank-you for you time.

Most respectfully yours,
Janis Hirsch

To Our First Lady, Jacqueline Kennedy,
    Now and Forever In Memory of our Late Great President
John F. Kennedy

.

<u>Never</u> (forget)
Dear Jacqueline Kennedy,
    Knowing how you feel with your Loss, I write in deepest regret and utter sorrow, sending my sincerest condolence to you, to your wonderful children, and the entire Kennedy Family. <u>Never</u> was there a greater President than our own, John F. Kennedy, to me he was a saviour of our troubled times and ills and was God given to us and the whole world.
    <u>Never</u> was such a vast job to be done left so unfinished as the one started by President John F. Kennedy.
    <u>Never</u> was there a greater Diplomat than Pres. John F. Kennedy, and Jacqueline you yourself his wife.
    <u>Never</u> was there a younger Pres. and First Lady as the Kennedy's.
    <u>Never</u> was there a more handsome President + First Lady as the Kennedys
    <u>Never</u> was there a more brilliant President + First Lady as the Kennedys
    <u>Never</u> in the history of the United States had a President met such an untimely death as John F. Kennedy
    <u>Never</u> will President John F. Kennedy be forgotten.

Never shall President John F. Kennedy be forgotten in our prayers.

Never could I come to an end of this Never letter saying all the nice things there are to say about our famous Pres. John F. Kennedy, so I will close by saying I cried openly for him and will Never be ashamed of it.

P.S. I will always try to remember President John F. Kennedy in my prayers, he has gone to a higher + greater achievement he has gone to God

Jacqueline, I am sorry for you in your loneliness.

John H. Jakusik

· ❧

Dear Jackie,

"Since I can't "Hello" to you
In person as I'd like to do,
I sent this special card to say
I'm thinking of you anyway."

I know you will never be in need of a friend, but if you ever want a new one to talk to I'm a very good listener.

You have been in my thoughts so much lately along with President Kennedy. I don't think you need fear that he will ever be forgotten. I'm just an average American—average mentality, average housewife, average housing, average size family, a year younger than you and perhaps a little more sensitive than some, but I will always have a warm spot in my heart for both of you as long as I live.

I live about 30 miles from New York City in a small town on Long Island. My telephone number is . . . [Number excerpted] if you ever want to talk.

Very fondly,
Marilyn Davenport

P.S. You probably will never even read this card, but if you do it is sent most sincerely

· ❧

DENVER, COLO.
DEC. 18, 1963

Dear Mrs. Kennedy,

I am writing for my brother and I trying to express how sorry we are for you, Caroline, & John

My Uncle Loyal Mull had a heartattack after he heard of the death of the Pres. and died 2 hrs later & now we know one thing for sure and that's that they are both in Heaven. My brother and I have decided we are going to earn about a dollar apiece and send it to Pres. Kennedy's Memorial Library

Merry Christmas

Linda Gayle & Donnie Nichols

P.S. I'm sending you a picture of Mary so don't [worry] she will take care of you.

NORRISTOWN, PA
DECEMBER 7, 1963

My Dear Mrs. Kennedy:

Today as I watched you walk into your new home, my prayers for your happiness and contentment went with you. Now you will enjoy a greater measure of privacy, where you can weep with your children, until all of your pent up emotions and the frustration of it all will slowly ebb like the surging tides. Take long tiring walks in the snow the wind and the rain, sing soft lullabyes to John-John at twilight before the lights are turned on in the evenings.

Allow John-John and Caroline to snuggle close in bed with you once in a while as a special treat to them.

Nothing heals wounds like the great outdoors and the warmth of little children.

I hope you will not resent the familiar tone of my remedy for sorrow and lonliness because every mother in America has held you with their arms of sympathy.

I am a negro mother and grandmother, poor in the material things of life, but rich from the thought that to-day and to-morrow, the American negro can breathe a little freer and hold his head a little higher, because of a great young man of courage. No more party criticism of plans and dreams well meant for we the common people, but a beautiful memory in the minds of all nations who walked with you from the North West gate in honor of his greatness and the compassion he had for all man kind.

I have for you two beautiful antique pieces of furniture if you will accept them, a lovely old marble top bureau and a graceful old sway back chair.

If you are ever in Norristown or are passing through you are most welcome to inspect them, to see if you would like them.

May the peace of the Yuletide be with you and the children.

My grandchildren, Andrea and Christopher Baptiste send Christmas greetings to Caroline and baby John-John

Most Sincerely

Ethel M. Robinson

....✻

Few expressions of grief and loss in the entire body of condolence mail are more poignant than those offered by children. With far fewer psychological resources to make sense of such a tragedy, many wrote with insight, eloquence, and compassion. Teenagers who had taken an interest in JFK's presidency wrote powerfully of the impact his life and death had on their coming-of-age. As one expressed it, "It is said you are not grown up until you have experienced the death of someone you love. On November twenty second I grew up." Large packets of letters from children sent from schools often clearly had their origins in a classroom assignment. Though they are touching to see, with their painstaking effort to print clearly, they have not been included in this collection. Occasionally a pupil strayed and offered his or her own spontaneous addendum to the task at hand. Their messages reward reading. Very often, however, children on their own wrote letters to Mrs. Kennedy from home. What follows are letters from children that capture their unique perspective.

Lisa Booth Blumberg
229 North Mountain Avenue
Montclair, New Jersey

Dear Mrs. Kennedy,

                                  empires
Suns will set, cities will fall,-emperour will rise but
the goodness of your husband will never beforgo$ten.

Of course there will be great presidents again, but There will
never be one like John Fitzgerald Kennedy.

Respectfully yours,

L. Blumberg

JAN. 17, 1964
LEVITTOWN, PENNA.

Dear Mrs. Kennedy,

Something vital is missing from the house on 1600 Pennsylvania Avenue. Maybe it is youthfulness, a lovely young couple, two charming children, or maybe a guiding hand, a gay wit, and a special security people drew from his presence, all these things and more, Americans lost when Lee Harvey Oswald committed his heinous crime. These losses are irreplacable.

Though you have lost a loving husband, Caroline and John a devoted father, Robert and his sisters and brother, a brother, Mr. & Mrs. Kennedy a son, and all Americans a cherished friend, we will not forget John Fitzgerald Kennedy. We share a wonderful memory though he died, still his ideals were not and will not ~~die~~ be buried. I promise you I will give body and soul to perpetuate the very ideals President Kennedy lived for. And, I am sure he would wish to be remembered for his humanitarian beliefs.

So now, in your time of grief, I offer to you and your children all I can, my deepest sympathy and a solemn promise for the future.

Sincerely yours,
Barbara Rimer (15 yrs. old)

....❀

MINNEAPOLIS, MINN.

Dear Mrs. Kennedy:

My name is Diana Tyler. I am enclosing a paper in which many Negro people have expressed what the death of President Kennedy has meant to them. I am very sorry that such a thing had to happen, it's as if history were repeating itself, in a way. I guess we needed that shock, but did it have to happen, especially that way? I'm not very good at expressing myself but all I can say is I sympathize with you. You have lost your husband, we have lost a president which has done a great deal for us. There is a poem which expresses the thought I felt when I thought of you and your family. The poem is <u>Lament</u> by "Edna St. Vincent Millay. The last line in the poem reads

    Life must go on;
    I forget just why

                  Sincerely yours,
                  Diana Tyler
                  14 yr. old

A FRIEND
WOODSIDE, CALIF.

Dear Mrs. Kennedy,

    I am only 9 1/2, but I nearly cried when I heard that shocking news. My little sister, Liz, was screaming, "My President, my President."

    He was such a <u>good</u> man and maybe the Good Lord thought he would be nice to have around at an early age, so he took him. If you want to send a letter to the President, send it to

        Care of the Good Lord
        President John F. Kennedy
        House of the Presidents
        Heaven

Send it by air Mail.
Keep this thought in mind.

                  Yours truly,
                  A sympathetic friend,
                  Kate Pond

1219 Chestnut
Orange, Calif.

Dear Mrs. Kennedy,

I hope the words I wrote last
year in fifth grade for Robert Frost
will express my feeling and provide
an epitaph for our late president.
  "The miles are gone,
   The promises were kept.
   I heard the news,
   I sat — I wept."

                    Sincerly
                    Douglas K. Grumblatt
                    age 11

Dear Mrs. Kennedy, and children,

I found a saying in a book about presidents, it goes like this: No king this man by the grace of God's intent;

No, something better,

Free man—President!

I think it fit your husband, President Kennedy, very well.

When I heard he was shot I couldn't believe it. When I heard it was true, I felt my own father had been shot.

I heard reporters say he's the second best president of the United States. To me he was the best.

My age is twelve. When I was eleven, just about a year ago, a shopping center had a sign up that read: Welcome Mr. President, for he spoke at San Diego State College. Now almost a year later it reads: Closed in memorian of John F. Kennedy.

> Your Friend,
> Jill Zarnowitz

DEAR MRS. KENNEDY

I AM SORRY HE IS DEAD

TONY DAVIS

Age 7

P.S. Tony wrote this the evening after hearing of the President's death and it so aptly and yet simply expressed the feelings of all of this family that I have decided to send it on to you.

> In Sympathy and sorrow
> Mrs. Donald K. Davis

[DECEMBER 19, 1963, POSTMARK]
LAKE ORION, MICHIGAN

Dear Mrs. Kennedy

I am very sorry your husband got shot, because you had to suffer so much, and I know it takes alot of brains to be selected President. I am only in third grade and we did not learn how to write. I know you should forgive your enemies, but it is hard to forgive Lee Oswald. It is hard to forget that the President got shot. I know you are receiving a lot of letters, my mother said you have about 2 million letters but try to get them read, because some are still more interesting than this one. You have many pictures in the papers, and they say you are very pretty, and I think so to, and your children are very cute. I wanted to phone you but my father told me not t,o because it is not polite, so I didn't. I hope some day we could make friends, I live in Lake Orion 31 mile road. If you would will you send me a picture of the President John F. Kennedy and you and John and Caroline, that is, if you have time. Some day if you would like to some day come over our house we live on . . . road, Michigan. We watched TV all day allmost, about President Kenedeys funaral. It was on day and night. It is very sad expeshely for you. He was the best Presidend in the whole world. Lee had a good weapon and he could not miss, I wish he did miss, and didn't even think of killing. I guess some people are that way, and don't think of what there doing. I bet you feel sad. Nobody will believe that we are writing to you, and we are not going to get a picture maby.

very respectfully,
Kevin Radell age 8

SHELTON, CONNECTICUT
SEPTEMBER 4, 1964

Dear Mrs. Kennedy,

Over the next hill about a mile away there is a monstrous Norway Spruce planted on the day of Mr. Lincoln's death. It is the one remaining tree of twelve that a man planted there a hundred years ago.

On that black day last November I asked my father for two blue spruce to plant in memory of your wonderful husband. Dad gave them to me and I planted them and they lived through their first winter and are growing fine. They are only about a foot tall now but I certainly hope they will grow forever because your husband did not save America he saved the world from being blown to bits.

How you had and have the courage to face life and the world is beyond me. I believe you are braver than any war hero and its too bad all people couldn't have your virtues.

I know I'll never forget your courage on that day last November and of the next few days. Even 50 years from now when I'm 64 I know I'll remember as clearly as I do this minute the shock, the grief and how I cried my eyes out and prayed for you and he.

Very Sincerely yours,
Sandi Jones

HOUSTON, TEXAS
DEC. 6, 1963

Dear Mrs. Kennedy,

I am ten years old. When I saw them moving President Kennedy's rocking chairs out of the White House, a great sadness entered my heart.

You made such a beautiful collection of treasures from other Presidents of the United States. Do you think you could find it in your heart to leave one of President Kennedy's rockers for children my age to remember him by.

I will always remember him laughing and talking with touring children and dignitaries alike, seated in his rocking chair, and most of all I remember his solemn and determined look as he sat pondering the terrible decision he must make in the Cuban crises.

Someday, I hope to be able to visit our beautiful White House and see all the wonderful treasures you have collected from Presidents of past generations, but please leave this one very personal treasure of a young president of our generation, so that one day I may run my hand over the wooden arms, and borrow for a moment a little of the courage and vigor of this our youthful 35th President of the United States of America.

May our Lord's Birthday and the New Year bring peace and comfort to you, Caroline, and little John.

Respectfully,
Suzan Elizabeth Lane

P.S. Please tell Caroline I wrote a poem for my English class about my horse, Silver, and her horse, Maccaroni.

1/24/64
HOPE, MICHIGAN

Dear Jackie.

I known Jackie will never read this, But I going to write it as thro she were reading it.

I known how you felt and still fell on that nightmare of a day. Because I felt as tho the world was coming to an end. That Friday I was in Gym when our gym teacher told us that the Gov. of Texas and President Kennedy were shot. I was like everyone else, I didn't believe her. I throught it was some kind of a joke. Then about 10 minutes later on the P.A. sistem they played the United Atum. I knew then that something awful had happened. Tears were filling my eyes as thro it were raining out. I ran in the lockerroom not wanting to know what had happen. Then my gym teacher came in and told me he had died. I didn't believe her. How could such a great man be dead. I loved him so much. I never wanted to die so much in my whole life. <u>I'm only 12"</u> but I loved" him so much. When he and nixon were running I prayed for him to win but he would have gotten it anyway. Buy the time I got home from school I eyes stunged so much it felt like a million needles poked in my eyes—I prayed to God Friday that when I'd wake up in the mouning I'd find it a terrible nighmare.

Saturday night I stayed up until 3:oclock that night watching people in the retunda going by his casket. Again I prayed it was only a night-mare. Than monday when I was watching his furnal," I got this sturp idea that it was only someone who looked a lot like him that got killed. He was such a great man. I or mostly anyone in the United States would have done anything for him. I still pray at night that he isnt really dead. But God can't even grant that. I ~~sill~~ still cry over him as thro he belonged to just me. He was a part of me.

Brenda Klemkosky

JANUARY 18, 1964
AUSTIN, TEXAS

Dear Mrs. Kennedy,

I know that you hate the whole state of Texas. I do to. I wish I lived in Washington, D.C. where maybe I could maybe see you standing on your porch. I am determined to move there as soon as I can. I would feel safer there. My greatest wish is to know you and to be exactly like you which will be hard. My friends Lorraine Atherton and Donna Sonnier and I talk, think and live you. We try to write your your name like you do, (Jaqueline Kennedy) or however you do it. We play a game in school. Lorraine is Bobby K, Donna is Teddy K., and I am you. Teddy (Donna) pulls out the chair to my desk, takes my jacket off for me, and opens the door for me. It's fun to be you. They say I look like you, too, although I am a blonde and wear glasses.

I love you more than anyone and I have an urge all the time to go to Washington to (maybe) see you. I love you an awful lot. I get every magazine + newspaper clipping or picture about or of you. I heard your speach twice on television. I think you are the bravest woman that ever has lived. I think that the Kennedys might be distant cousins of mine 1. because my grandmother told me they are. 2 because my grandmother's great—grandparents came from Ireland and were named Kennedy 3, because I still have cousins all over the place named Kennedy 4. because they look like the Kennedy men.

I think I am going to the World's Fair this summer and we might get to drive through Washington (I have cousins there) I if we do I might get to see your house.

I know you couldn't care less about me, but I love you. More than anything I do. Don't you understand? I LOVE YOU! PLEASE! believe me,. (If you get this letter) (I know its not ESPESIALLY TOUCHING).

> I Love You,
> Jane
> Dryden

....❊

SOUTHAMPTON, N.Y.
[NOVEMBER 27, 1963, POSTMARK]

Dear Mrs. Kennedy,

I'm so sorry that people are so full of hate that they had to kill Mr. Kennedy. He was a good president.

He can see you from heaven and when he knows that you're unhappy he'll be unhappy too, but he is with you realy all the time with God. When you love each other, you never realy part.

We had two kittens before one died.

One night I was looking at a pile of blankets and on top of them I saw the kitten that had died, sleeping there. It may have been just the shape of the blankets, but I think that God let her come down and see me.

Love,
Ellen Junker

DEC. 13, 1963
LEBANON JUNCTION, KENTUCKY

Dear Mrs. Kennedy,

On November 22, 1963 the United States and the free world lost a great leader and a great man. Even though I have never met him he seemed close to all of us.

On November 22 it was a friday all the kids where anixous to get home because friday was a day when you could stay up as long as you wanted to sleep late and no homework. We all said T.G.I.F. meaning "thank goodness its friday. One girl was so anixous to get home she pushed half way down the stairs.

When I came out the door a girl was waiting for us to come out then they all rushed up to us and said the President was shot and he is dead. We couldn't believe it we thought it was a joke but no one was laughing. . . . When I got home the television was on I took my coat went hang it close that closet door and hit my fist on the door and laid on the bed and cried next I got my diary and wrote President Kennedy has been shot he is dead. What in the world has happened when people try to make this world a better place to live in and some one shoots the president. I can't write or fill what I would like to say. But God bless the Kennedy family.

When I heard that they think they had found the man who shot President Kennedy Lee Harvey Oswald I felt no love or hate toward him

On Sunday I heard that Oswald had been shot and killed I hated him and I said he had a crime to pay for he got off easy. Mrs. Kennedy when you walking behind behind the cassions you look like a general leading an army into battle you had more medals than Highly Selasse. You stood taller than Charles Degaulle. . . .

We will always remeber Mr. Kennedy for what he has done for the Negroes, jobs, education phiscal fitness and peace. And so Mrs. Kennedy I close my letter hoping you can read it. May God bless you in what ever you do and all the free world hearts go out to you and the Kennedy family.

> Sincerely yours,
> Teresa Bradbury
> Lebanon Junction
> Kentucky

BROOKLYN, N.Y.
NOVEMBER 27, 1963

Dear Mrs. Kennedy,

I am very sorry about the death of President Kennedy. I know he did much to help our country and to show how much I liked him, I am going to try to be a better boy.

All the children in my class are praying every day for the President and for the family. I too am praying very hard especially for you and Caroline and John.

May God Bless you.

> Respectfully yours,
> George Wysota

Dear Mrs Kennedy,

Since November 22, I have shared your grief for the lose of our beloved president, but only now can I bring myself to write it down and tell you.

I live in a small town (7,600) where I am a freshman in high school. Batavia is very Republican. When Mr Kennedy came through Batavia in 1960, he was met with Nixon posters. Though I was only in sixth grade, I managed to push my way through the crowds to shake his hand. It was the thrill of my life and I shall never forget it.

I was taking an Algebra test on November 22 when our assistant principle came in to talk to our Algebra teacher. He left and our teacher turned around and said the words which echo in my mind—"The President has been shot—in Texas. It could be fatal." People looked at me, but I didn't cry. "He's all right" I kept on saying and I prayed. Algebra was over and I went to Social Science. My best friend met me there and said, "Be, he's dead." I didn't say anything. I just kept thinking it isn't true. Miss Sieler came in and told us it was true. He was dead.

Now Social Science is a discussion class and about half the time, it's polatics. I'm outnumbered 30-1. Teacher and all. People used to say, "Kennedy's doing this or that wrong" and I'd say "He is not! What would you do?" I always had an answer because I don't think I missed more than three press conferences and I still have my scrapebook from 1960 with over 300 pictures of you and him, so I could quote about anything he said. Well, now he was gone. Everyone looked at me. There I sat—The Big Democrat, The Kennedy Defender (Nick-names I was proud of). Now he's gone. How will she take it? They seemed to say.

I tried not to cry, Mrs. Kennedy, I really tried. During the funeral and the burial I tried, but I just sat in front of the television and thought of him, and you and Caroline and John-John and cried and I'm crying now.

We had a service for him that Sunday. I played in the Band. All I can remember is not being able to see the music because I was crying. That's all I can remember. He was a great man and a Republican town paid tribute.

I could ramble on for hours but I won't. I can not say anything to ease the hurt. If I could, I would say it to myself, too. But, though I didn't even know him, I loved him. We must think of him in heaven, for he is happy.

Yours very truly,
Bridget Tierney

··❧

After leaving the White House, Jacqueline Kennedy returned to private life. The days of tumultuous political life in the nation's capital, glittering state dinners, and the constant duties of First Lady were behind her. She remained one of the most photographed and publicly pursued women in the world until her death in May 1994. But she sought throughout her life to guard her privacy. It is striking to see how many Americans recognized from the moment of the assassination that they had lost not only the President but the First Lady and children they felt they knew so intimately. The loss of the vicarious pleasure many took in the young White House family itself was a source of grief. Some asked Jacqueline Kennedy "not to leave us." Many thanked her for her service to the country and offered this valedictory: "Jackie, we shall miss you."

CHICAGO, ILLINOIS

My dear Mrs. Kennedy:

I just saw your lovely face on television and heard you speak. And all the while, a feeling of great pain overwhelmed me because I know that your public appearances will be all too few and the family we have come to love so dearly will become news we find occasionally in the papers. Yes, we have come to think of all of you as belonging to us. It was so wonderful to feel part of the great excitement our dear president gave us all. And to be part of you, watching the children grow and seeing the grace and beauty both of you brought to our land. Yes, a lovely bright light went out for us, and we share with you and the children and the family the tragic loss that can never become a thing of objectivity. But we have another loss—we have lost you and the children and we miss you very much. We felt that we could watch Caroline and John-John grow, and share with you and our beloved president the joy of family. And now we feel doubly bereft.

I know you have received so many messages very much like this one, but I did want to tell you how lonely we are for the sight of our First Family. The brightness and electric excitement have gone out of the news, for we cannot get used to turning on the television and not seeing that wonderful smile and the brilliant eyes. We could tell when a wry remark was coming and watched for it in delighted anticipation. It isn't the same anymore. We will back everything he fought for so hard, because of him, but there is no pleasure behind it. I do not mean any disrespect to Mr. Johnson for I know how hard it must be for him to even begin to fill a role that such a brilliant, sparkling man left vacant. But ordinary men seem to be in government now. And government will go back to being ordinary once more.

Please don't disappear from our view. We want to know how you are faring, how the children are, if you all are well, and above all we want to have you lean back and rest against us, knowing that our love does sustain you and is there for you and the children, dear Mrs. Kennedy. We know how great was your loss because our loss was great too. How we love him; I shall not say loved, because his memory will not dim for us. He was a lovely shining knight and we are thankful that we were privileged to know him and to have him lead us.

May I write to you from time to time to find out if you are all right and if the children are all right too? We love you very much and don't want to lose touch. We hope to be able to come to Washington soon to visit "our grave" and pay our respects.

Keep well and God watch over you and the children.

With deepest affection,

Ethel Bedsow

Ethel (and Norman) Bedsow

JAMAICA, N.Y.
NOVEMBER 26TH 1963.

Dear Mrs Kennedy,

You will remember some time ago, I mailed a letter to your Dear Husband, in care of you. He received it and I got a reply. I am that same person (Picture above).

This is the saddest moment of my life, For never did I ever think such a tragedy would have happened to you and a President so many of us loved. The first time He came on the Air to talk to us, I said to my Husband, That's my Boy. and so I knew He was the man these United States was waiting for, Now, I wish he had lost, But we cannot look into the future.

It is true that we all say "God knows Best but I will never believe, God was instrumental in this Destructive tragedy For God is a Good God.

Never in the History of these United States, such a Demonstration of Love and Affection has been shown to any President, even Mr Roosevelt who I had loved as much as your Husband. I watched every minute that My T V showed. I cried with you, I grieved with You, I walked with you, I even knelt at the Altar with you, I stood beside you at the Cemetery as the flag was placed in your hand, My spirit will always be with you, for you like your Dear Husband are wonderful.

Many critisized him when they saw little John John in Arlington with his father, But I thought it was a very adorable action on his part, tho' I know he had not planned it, To me it was lovely seeing him toddling around the Daddy he loved, Now my heart also goes out to him.

I know that God in His Mercies will keep close to you and your children, for I do think you are, and will be a devoted Mother to them. I don't care after this who will be a first Lady, but you will always be that to me.

It is said that time will heal all wounds, but Will it? For I have been in similar positions and my heart still aches for my loved ones,

The cruel death Our late President met, we will never forget, as he was taken in the bloom of his health and life. I don't want to sadden you, I want you to know that there are many of us who will always be devoted to his Memory, as a Good Clean faithful President. He will be missed by each American Democrat. Those who hated him are Nonenties, and are in the minority, so never let that disturb you.

To his Mother and Father, My heart goes out to them also knowing it was their love that made him what he was. So may the Good lord strengthen them to bear this cruelty that been dealt out them, for they too are wonderful Parents.

Take care of you babies, I know their fathers Spiritwill always be near to guide them, and that will be my constant Prayer, May he give you the strength to hold your head up high as you always were, Do Never forget God is a Good God and He will never forsake us who loves Him and keep close to Him. I would like to know where you will be, so I can always drop you a card of Cheer

Sincerely & Faithfully

M.R. McCormack

(Mrs) Melbro R. McCormack.

....❧

DEC. 5, 1963
SPRING LAKE HEIGHTS,
NEW JERSEY

Dear Mrs. Kennedy,

Our nations loss. You not being in the White House. You and the President were the ideal couple. He was so loved, by all people—even me, a staunch Republican, and now when ever I hear some thing I don't like, I immediately come to your rescue. So gracious a First Lady. So many wonderful ideas for the White House—you know it was startling at first when you read about one of your projects, but then, you realized—what a wonderful idea, And that tour of the White House—I lived every minute of it, not even seeing it in person—it was the next best thing. The American people thank you for making people like me aware of what a wonderful country this is. There you were sharing it with the World—It wasn't yours, you were just taking care of it for us, and you wanted us to know. I'm almost afraid—me—writing to you—but our town misses you. When this tragedy struck, I felt like Peter Pan—when Tinker Bell was dying—if the children clapped long enough and hard enough—Tinker Bell would hear them and come alive—Well, they did clap, and she did live. And I thought if the American people cried hard enough and long enough—We would wake up and it would all be a dream—Well the American people did, and still are—but this isnt a fairy tale, this is for real, and we are all a part of it. I want to say "thanks," for knowing you and your husband—laughing with the sense of humor—crying with your sorrow—enjoying the children—Saluting the flag now is no more sing song—Every word is like a word from my prayer book.

To you Mrs Kennedy—Carry on as nobly as you did during those long four days—We walked every step with you—you made the American people so proud. We were so fortunate to have you in the White House—your ideas will live for ever. May the days ahead lessen the pain—may you find peace with God—and forgiveness in your heart. Good things will come from you. How do you end a letter to a "First Lady"—? I don't know—I just know your wonderful and thank you for a job well done. The country will miss you.

Estelle Sherman
(Mrs. George E.)

Dear Mrs. Kennedy:

You know the admiration and respect—the devotion—I had for your husband. Now you should know that in addition to all the sympathy in the world, this universal admiration and respect is now yours. One sunny afternoon, on the eve of the election, holding Caroline by the hand, the Senator walked toward his campaign plane, and I sensed he was walking into history. You have walked with him. He came back, a few days later, the President. You became in every sense the First Lady. That is no help to you now, but it may help a little to know that you have the nation's deepest thanks.

Sincerely,

Harris Wofford

The care and emotion in these condolence letters reflect the reactions of an extraordinarily diverse collection of American citizens. Not long after the assassination, two writers commented explicitly on what they believed would be the lasting significance of the hundreds of thousands of letters they knew Mrs. Kennedy was receiving. It seems fitting to end this book with the observations of one who explains why she had taken the time to write a letter, knowing full well the enormous volume of messages already delivered. The other offers a raison d'être for a collection of letters that preserves the response of ordinary citizens to the death of President Kennedy.

THOUSAND OAKS, CALIFORNIA
JANUARY 19, 1964

Dear Mrs. Kennedy:

There is only one way to be absolutely sure that in future generations no person can say, "Eight hundred thousand? Well, that's really not very many out of some one hundred and seventy million, is it?" At least one letter will represent the uncountable numbers who never express their thoughts and feelings through correspondence to editorial pages, or to their representatives in government, or to people not personally known to them.

As a registered Republican who in fact is an independent, as a voluble anti-Kennedy-dynasty citizen, and as a voter who disapproved of many, not all, but many of the late President's programs, his death shocked me more than words can encompass. Yet the depth of my grief was almost as large a shock. To be robbed of his presence on this planet, of the privilege of voting against him; many of the silent ones in this republic, this very great republic, experienced thoughts and feelings they did not and probably never will communicate to you.

It is my belief that this percentage of Americans may be the highest of all. In any event, they are now a part of that museum.

I was so proud of that man, of you, and of the family. So proud, but so shocked that I never even knew it until a sick, confused soul destroyed one of those who sought to help him most.

My very deepest sympathy to you and your children during this bleak period of adjustment and activity.

Very truly yours,

Sheila J. Lynch

....❧

FRESNO, CALIFORNIA
DECEMBER 6, 1963

Dear Mrs. Kennedy,

This is one of about 300,000 letters which you will receive from people who love the memory of your husband and who love you. I realize, of course, that you would find reading them all insupportable, but I write, knowing that you will be told again that many, many of us are in your debt, and would acknowledge it.

For several years I have studied you casually and have tried to be led by your exemplary behavior. I speak a little softer now; and, you have reinforced and refined my convictions about parent-child relationships. On November 25th you set another standard for me to seek. Watching you, I stopped crying, and I believe you have made me a gift of greater self-control in adversity. So far, I have only to remember your courage to take a share of it. Thank you.

There will be letters among the many you are receiving which will have merit for a collection to Mr. Kennedy's memory. I would urge that someone with talent (Mr. Sorensen?) edit them for publication with an eye to choosing excellent letters representative of these expressions: Mr. Kennedy's personal impact on us (there will be thousands that say "I feel I have lost someone from my own family"—and there will be a few that say it beautifully); faith in resurrection; recognition of his achievements in government; and, phrases which actually have the power to console—not maudlin, but magnificent.

There will be letters among the 300,000 from other persons, who have been widowed or bereaved, who will have bestowed their own endurance, faith and courage on you in an effort to carry you through this time. Those letters can be worked into a classic chapter, almost prophetic, which would help all people sustain their integrity during grief.

My parents are aging: everyone I love is mortal. I want such a chapter for my own. Would you share the letters with us all? Would you give us the book?

I don't know what to wish for you except—God bless—

Sincerely,

Grace Longeneker

Acknowledgments

Authors often dream about writing acknowledgments—the pleasure that awaits the completion of an intense project. But never have I so looked forward to expressing my gratitude than in the case of this book. My debts to those who helped make it possible are legion. First thanks go to the extraordinary letter writers and their heirs who responded to requests for permission to publish with such heartwarming generosity. Their kindness, their memories of their loved ones, of President Kennedy, of their lives, and of the state of the country in 1963 inform every page of this book. In only a very few instances did those who were approached decline to give permission to publish. All had good reasons, which, of course, have been honored. But it is revealing that none of those reasons involved a change in the sentiment expressed in their, or their family member's, letter. Thank you to one and all for making it possible to bring these letters to the public. The time the letter writers took to write to Jacqueline Kennedy not only reflects their personal thoughtfulness and generosity, it preserves for future generations an important part of our history.

My second debt of absolutely enormous proportions goes to the tireless, imaginative, dedicated, and oh so patient team of permissions researchers who found the letter writers—Mary Dalton Hoffman, Sarah Thorson Little, Ellen Lohman, Ann Louise Rossi, and Josh Sucher. Sarah Miller-Davenport's excellent sleuthing through court records in Chicago at the eleventh hour made her an honorary member of the group. The willingness of all these individuals to undertake a seemingly impossible task made the book a reality. Without genealogist extraordinaire Sarah Thorson Little there would have been no *Letters to Jackie*. The fact that I found Sarah through letter writer Ann Lounsbery Owens only deepened the sense that we were all in it together—thank you, Ann. Mary Dalton Hoffman led us all with her amazingly cheerful and resolute determination to find the letter writers nearly a half century after they penned their messages. Her sense of humor, good judgment, and support kept me—and the whole permissions group—going on many occasions. I gratefully acknowledge as well the valuable assistance of the New England Historic Genealogical Society and genealogist Polly FitzGerald Kimmitt. The staff at the John F. Kennedy Library proved unfailingly helpful and gracious with my unceasing requests for assistance. Thank you to the Library's director, Tom Putnam, for his early encouragement and to Chief Archivist Allan Goodrich for granting me access to unprocessed materials as well as sharing his knowledge of the collection and all things related to President Kennedy. Archivists Sharon Kelly, Stephen Plotkin, and Michael Desmond not only found all the material I was looking for, they kept this scholar steady company in the beautiful research room at the library overlooking Dorchester Bay during the time I spent there reading thousands of condolence letters. Laurie Austin's assistance in locating photographs uncovered many of the wonderful images in the book. The dedication and professionalism of all made research at the John F. Kennedy Library a pleasure for this historian.

To Lauren Dinger, my devoted research assistant, and to Elizabeth Armstrong who recommended Lauren, my deep appreciation. Whether through photo research, tracking down citations, proofing letters, organizing thousands of archival documents, contacting letter writers, and an endless array of additional tasks, Lauren's efforts shaped and enriched this study. I thank her for all that she did and for being so good-natured while she did it. Leslie Hendrickson and Kelly Gu did yeoman's work early in the project that provided a solid footing for the book. Genevieve Smith contributed her superb fact-checking skills late in the process. I am very grateful for their efforts.

Bette White and Lynda Gaudiana undertook, with utter selflessness, the very difficult task of transcribing the letters. Their skill, interest in the letters and the writers, commitment, and unflagging support—all of it exceeding even the most capacious bounds of a nearly forty-year shared friendship—made it possible to complete the manuscript. It's hard to imagine the book without them and without Russ's good cheer as value added.

Nancy Tuckerman was very gracious about answering my questions. Her observations and memories added substance and texture to the story. I thank her very sincerely for her willingness to speak with me about the condolence letters.

Several historians provided excellent advice as I prepared this study. Robert Dallek offered encouragement at the start of the project. Robert Clark, supervising archivist at the Franklin D. Roosevelt Library, answered crucial questions. I benefited enormously from the comments on Kennedy, civil rights, and race offered by several outstanding scholars of African American and/or Southern history—Jacqueline Jones, J. William Harris, John Dittmer, Catherine Clinton, and Vanessa Northington Gamble. Bill Harris read not one but three drafts of some material, improving each one with his suggestions. Jim Lehrer offered valuable assistance as I explored events in Dallas.

Much of the work on this book was completed during the year I spent at the Radcliffe Institute for Advanced Study. To the staff and fellows of the Institute, my very sincere thanks for your assistance, intellectual energy, and delightful company. To Judy Vichniac, director of the fellowship program, a special thanks for friendship, support, and generosity. Emma Rothschild's encouragement also meant a great deal to me. The "book crisis group"— Judy Coffin, Susan Faludi, Willy Forbath, and Russ Rymer—made the year at Radcliffe so special. Our evenings together discussing our work combined fun—hilarity!—and great conversation about books and history. Thank you all for your friendship, and those memorable evenings. To Gail Mazur, brilliant poet and wordsmith, as well as dear friend, enduring gratitude for reading sections of the manuscript and offering so many superb insights and observations. Sara Rimer, Kit and Jane Reed, and Ellen Rothman urged me to undertake a history that might reach a wider public—I'm grateful for their frequent infusions of courage.

The entire Ecco group has earned my admiration as well as enormous gratitude for their commitment and incredible skill at every stage of the process. Thanks to Allison Saltzman, Rachel Bressler, Greg Mortimer, Michael McKenzie, Rebecca Urbelis, Mary Austin Speaker, John Jusino, Doug Jones, Carla Clifford, Kate Pereira, Jeanette Zwart, and the remarkably efficient and dedicated Abigail Holstein. Dan Halpern's commitment to the project made it happen, as did, most of all, my editor, the amazing Lee Boudreaux. Lee somehow manages to improve a book with such a deft touch that the author hardly senses the transforming alterations! Her commitment to and enthusiasm for the book, as well as her graciousness, have sustained me. It's been a pleasure to work with Lee and the Ecco staff.

At so many points along the way, this was a project that might not have been. But the person who opened the door and kept it ajar was Lindy Hess, who understood the idea, valued its purpose, and remained steadfast and enthusiastic as I carried out the project. Among her many selfless and generous acts was an introduction to Scott Moyers. Scott's belief in the book, his wisdom, warmth, and brilliant advice shaped all that followed. Revan Miles helped me to understand better what I was reading in the condolence letters—a gift I can only acknowledge but never adequately describe.

My family and friends listened patiently as I made my way through these remarkable letters, and, along with Winslow, kept me company through the writing. (My mother and my sister Maureen also offered sage advice about the project.) I thank them for being by my side through this and all my other life adventures.

The Letter Writers

The letter writers or their heirs were invited to provide a brief biographical statement about the letter writer. Some wished to remain largely anonymous, while others offered varying material and observations. Their remarks form the basis for the information offered below. Data on some deceased letter writers has been acquired through genealogical research in public records.

Albert, Mrs. Clytus
Anna Lou Albert was born and raised in Breaux Bridge, Louisiana, one of eight children. A devout Catholic, she was the mother of six children, and at the time of her death, she was a grandmother to ten. Although she had only limited education, Mrs. Albert wrote letters on all sorts of occasions. She was an excellent seamstress. Her daughter recalls that her mother "loved helping people and loved visiting people . . . she was a giver, a healer, and touched everyone that she came in contact with." Mrs. Albert died at the age of sixty-nine from congestive heart failure.

Alley, Miller A.
Miller Alley's son reports that his father "was sincere, very sincere with regard to the letter. He passed away in 1972 of a brain tumor. Dad loved to go fishing with his buddies."

Andersen, Sue Ann
Sue Ann Anderson finished her university education in early child development and taught kindergarten in both private and public schools. She joined Wycliffe Bible Translators in 1984 and has worked in Papua New Guinea as a literacy consultant as well as facilitator for two New Testament translations.

Anderson, Susie

Susie Anderson had a successful career in accounting and is now semiretired in Arizona. She is the mother of a son. She still vividly recalls meeting John F. Kennedy.

Anonymous

This letter, addressed to "John-John" Kennedy, was unsigned and provided no return address.

Archer, Mrs. B. F. (Pearl)

Pearl Archer lived from April 1891 to September 1979. She was a homemaker who enjoyed baking pies and was active in the Eastern Star fraternal organization. Her son is now ninety-one and lives in a nursing home. He remembers clearly that his father, who worked for the railroad, died the same day and hour as John F. Kennedy.

Ashburn, Nancy

Nancy Ashburn graduated from high school in Beacon, New York. A talented musician who played the piano and the viola, she also attended nursing school and studied ballet. She worked in a variety of settings, including as a construction equipment operator for the New York Thruway Authority. In that capacity, she was a member of the rescue/recovery team sent to the World Trade Center after the terrorist attack on September 11, 2001. She died in March 2007 at the age of fifty-seven.

Asselta, Madge E.

Madge Asselta died of breast cancer in 1978. Her daughter recalls: "I was seventeen years old at the time of the assassination and I remember clearly the sadness in our home."

Baines, Mrs. Doris

In 1967, Doris Baines and her family moved to Westover, Maryland. She was a wonderful cook who, her family reports, "became famous for her pastries. Everyone knew her as the 'Cake Lady.'" Mrs. Baines is now eighty-one years old and resides in a long-term-care facility in Princess Anne, Maryland.

Barnes, Ricco

Ricco Barnes was incarcerated for armed robbery when he wrote to Mrs. Kennedy from the Cook Country Jail. His father served time in a state prison shortly after Ricco's birth. Ricco lived a long life and died in 2008.

Becker, Mrs. Duane

Mrs. Becker and her husband farmed from 1942, the year they married, until 1977, when they retired and moved to Mondovi, Wisconsin. They were active Catholics and enjoyed fishing and playing cards. After her husband's death in 2003, Mrs. Becker moved to a nursing home, where she lives at the present time.

Bedsow, Ethel

Ethel Bedsow was born in Saskatoon, Canada. She grew up in New York and attended college. She settled in Chicago after the war and worked with prisoners, helping them to further their educations. She instilled her love of literature in her son and stressed to him the importance of a good education. She loved books, played piano by ear, wrote poems, and had, her son observes, "the gift of sentimental prose, and she meant every word she wrote in her letter" to Mrs. Kennedy. She returned to New York to care for her mother and died there in 1987.

Bentley, Mary W.

Mary Ward Bentley was born into one very influential Rochester family and married into another. Her father-in-law, Sardius Delancey Bentley, was a prominent Rochester attorney, and both the Ward and Bentley families contributed to education, science, and civil rights initiatives in Rochester. Her great-niece Eleanor remembers her "as a very kind, gentle woman, who had great empathy for those around her." Mary W. Bentley died in Rochester in February 1973.

Berkery, Margaret

Born in 1917, Margaret K. Berkery was interested in art, opera, and travel. She had a large cane collection that featured animal heads and was also known for her sense of fashion. She loved New York City and was devastated after the 9/11 terrorist attacks; she spent much of her professional life, until her retirement in 1982, working as an employment supervisor for a company located in the World Trade Center. She died in 2005.

Bethell, Thomas N.

Thomas N. Bethell writes that he "was a book editor in Boston when he wrote to Mrs. Kennedy." In the mid-1960s he moved to Appalachia to work in the War on Poverty as a supervisor of VISTA volunteers and then as a reporter for the *Mountain Eagle*, a crusading Kentucky weekly newspaper (where he is still a contributing editor). During the 1970s he served as research director for the United Mine Workers of America. For the past thirty years he has been a self-employed writer-editor, policy analyst, and consultant in Washington, D.C. He is married to Katharine J. Bethell, a retired social worker, has two adult stepchildren, and, as of November 19, 2009, one grandson. Despite the passage of time, he remembers November 22, 1963, as if it were yesterday: "the pain and sorrow never really end."

Bingle, Mrs. Charles

Mrs. Charles Bingle's admiration of and loss of her own father, her daughter notes, strongly shaped Mrs. Bingle's response to the death of President Kennedy. Her mother identified with Mrs. Kennedy. Her mother was a letter writer who followed events on television and then was often inspired to offer her own commentary. Mrs. Bingle died in 1983.

Blumberg, Lisa

Lisa Blumberg attended Wellesley College, where she majored in political science, and later went on to study at Harvard Law School. In addition to her career as a corporate counsel, she has been actively involved in the disability rights movement and has published essays on the subject. She recalls that "as an elementary school student in the early 1960s, I was inspired by President Kennedy to believe that anything was possible."

Boling, Patricia Lee Rita Mary

Patti Boling has been married for thirty-seven years. She is the mother of three children and grandmother to five. She returned to work when her youngest child was in fourth grade and lives in central Ohio.

Bond, Beatrice Joan

Mrs. Bond was a deeply religious woman who enjoyed reading and letter writing. She died in 1985. Her daughter remembers her as the "epitome of kindness, caring, and compassion toward her fellow human beings."

Boner, Mrs. Linnie

Linnie Boner's daughter observes: "My mother was one of the old-fashioned ladies of the past. Probably a lot closer to the Proverbs woman than any I have ever known. She passed away at the end of December 1999, just a few days short of seeing the new millennium, having turned ninety on November 3 that year."

Boorey, Mrs. Natalie

Natalie Boorey enjoyed travel with her husband in her later years. She is deceased.

Borders, Mrs. Frank

Mrs. Borders's daughter describes her mother as "a private outspoken lady who loved God, her garden, her family, and her country. She lived an unassuming life with few excitements other than births, deaths, and graduations. She didn't do anything extraordinary except die of gangrene in 1993 and get her first anything published seventeen years later." She notes that her mother "was very fond of President Kennedy and thought he could have done great things for the country if he had lived. She also thought that is why he was killed."

Bradbury, Teresa

Teresa Bradbury was thirteen years old when she wrote to Mrs. Kennedy. She has one child and lives in a small town in Kentucky.

Bradley, Mrs. Joseph, Jr.

Veronica Bradley raised five children. She retained her admiration for John F. Kennedy and considered him to be "the best president we ever had." She died in 2003 at the age of seventy-two.

Brenner, Peter
Peter Brenner died just five years after he wrote to Mrs. Kennedy. He had three daughters and seven grandchildren who recall "his heroism in World War II" as well as their "father at the manual Smith-Corona typewriter, fragrant pipe smoke wafting through the air, classical albums on the RCA, pouring his heart and soul into a letter that simply had to be written."

Brooks, Mrs. Wilma
Wilma Brooks was born in Kentucky in June 1930. She grew up in Indianapolis and was married and divorced at a young age. She worked in various restaurants and was known for her pecan cinnamon rolls during her years of employment at a delicatessen in Indianapolis. She enjoyed cooking, worked very hard her entire life, enjoyed reading, and was close to her family. Although she had no children of her own, her sister notes that "she adored my children and put claim on them." Raised a Baptist, she converted to Catholicism. Her family reports that she died in 1980 at the age of fifty on a hot afternoon, as she was in the middle of moving to a new apartment in Indianapolis. She was pulling her possessions in a wagon; the heat contributed to her death. Her mother and sister survive her.

Buerman, Fred
Fred Buerman survived to the age of ninety-six. He lived with his extended family and left a lasting impression on several generations. His granddaughter is now eighty-four years old. Mr. Buerman's great-granddaughter remembers him "in his wheelchair with handlebar mustache and cigar. He would tell me stories of crossing the Mississippi River when it was frozen in a covered wagon."

Burke, Roger L., Jr.
Roger Burke worked as a dishwasher at the Park Street Diner in Ayer, Massachusetts. The owner's son remembers him as suffering from a physical disability that affected his speech. He was very bright and was an avid fan of the Boston Red Sox. He moved away from Ayer when the diner was closed. He died in 1998 at the age of fifty-four in a church-sponsored home for those unable to find affordable housing.

Burril, Mrs. Andrew
Bertha Burril suffered from a severe heart condition from childhood and was not expected to live to adulthood. In fact, she survived to the age of seventy-six and had two daughters and two grandsons. Her grandson recalls that though she had only a third-grade education, she greatly impressed upon him the importance of education. She is remembered as a "good Christian lady" who "loved life" and was much adored by all who knew her.

Cannon, Irene

Mrs. Cannon lived twenty years after her letter to Mrs. Kennedy, dying at the age of eighty-two. Her son Joel, who became a real estate broker in Washington, D.C., reports that "she passed on her considerable writing skills" to his brother "Lou and his son, Carl, both of whom are successful and well-known journalists."

Carriker, C. O.

Born in 1895, Charles Omer Carriker worked as a farmer and machine shop tool and die operator; later in life he owned and managed property. He was retired when he wrote to Mrs. Kennedy and was diagnosed with Parkinson's disease not long after. Although he had only attended school until the fifth grade, he was self-taught and wrote a book on his life. Mr. Carriker died in March 1980.

Cates, Robert T.

Robert T. Cates, M.D., was born in Mississippi in 1927. He graduated from Mississippi State and went on to earn his medical degree from Tulane. He completed an internship at Charity Hospital in New Orleans and his residency at Baptist Hospital in Jackson. His daughter recalls that her father was a Mississippi family doctor who practiced medicine "the old way and was a brilliant man." Dr. Cates enjoyed fly-fishing and duck hunting. He had two daughters, one of whom became a physician. He died in 1995.

Chapa, Minerva

Minerva Chapa Medrano became a highly regarded and much-honored elementary school teacher. Scolded when she first attended school for not knowing English, she went on to graduate from high school and to earn a bachelor's degree from Defiance College and an M.A. in education. She died at the age of forty-four in 1993, leaving a husband, two daughters, her mother, and eleven siblings.

Chapman, Marjorie B. "Peg"

Marjorie Chapman came to the United States from London, England, in 1911, when she was eight years old. Her father was an Irish immigrant. She became a teacher and taught for nearly forty years. Her husband was also a teacher. Married for nearly sixty years, Mrs. Chapman and her spouse died within seventeen days of each other in 1985.

Charley, Pearl

Pearl A. Charley was born in Yakima County, Washington. She is an enrolled member of the Klickitat Band of the Yakama Indian Tribe. Her deceased brother, Leo Aleck, was the last medicine man for the Yakama tribe. She is eighty-three and resides in Okanogan County, Washington.

Clarke, Mr. and Mrs. John W.
Lynne Clarke, who wrote to Mrs. Kennedy also on behalf of her husband, was a Phi Beta Kappa graduate of Douglass College. An accomplished writer, she had a long career in government. She served as director of economic development for Monroe County in Rochester, New York. Mrs. Clarke died of breast cancer in 1996. John Clarke writes: "At the time the letter to Mrs. Kennedy was written, I was a radio newsman at WHAM in Rochester, New York, and coincidentally had aired the news bulletin reporting that President Kennedy had been shot in Dallas. I was graduated from the Cornell Law School in 1967 and have been a trial lawyer for over forty years. My areas of practice are environmental, health, and commercial law." He has been married to his wife, Stephanie, for ten years.

Cohan, Mrs. Donald S.
Mrs. Cohan is now seventy years old. She has three sons and five grandchildren. She continues, in her words, her "love of music, history, Civil War, all sciences." She has "bred golden retrievers and still remains anti-war."

Cohee, Mr. and Mrs. Ralph
Mrs. Cohee and her family struggled for survival after 1963. Despite her own hardships and health problems, she focused on the importance of educating her children. She enjoyed fishing, farming, and churchgoing.

Cook, Mrs. John G and Mrs. Henry Wood
Born in Georgia in 1917, Henrietta Cook was one of seven children. She had an eighth-grade education but enjoyed learning. After her marriage to John Cook, she devoted herself to raising her son and to being a homemaker. She was a wonderful cook and canned her own fruits and vegetables. She died of cancer at the age of fifty. Her mother, Mrs. Henry Wood, was born Cordelia Hearn in August 1884 and outlived her daughter. She was a lifelong Georgia resident, living on her own at the age of ninety. She had a limited education but enjoyed television in her later years. She also was an expert quilter, especially accomplished at tatting, or lace making.

Crabtree, Janice R.
Janice Crabtree lives in Texas. She is a mother and grandmother who was widowed and who remarried again in her eighties. She vividly remembers the Kennedy assassination, noting that it had a powerful effect on her long after November 22, 1963.

Cuchia, Mr. and Mrs. Frank
Jean Cuchia was born in Kansas City, Missouri, in 1924. Her parents were Italian immigrants who settled first in New York. Mrs. Cuchia and her family moved to Dallas in 1941. She married, had two sons, and today still lives in Dallas.

Daniel, W. J. B.

Mr. Daniel worked as a radar technician during World War II. His family reports that he developed skin cancer, secondary to radiation exposure, and was nearly bed-ridden at the time he wrote to Mrs. Kennedy. He had five daughters and died in 1973.

Davenport, Marilyn

Marilyn Davenport raised four children as a "stay-at-home mom for almost twenty years" before returning to work. Widowed in 2006, she and her husband were married for nearly fifty years. Mrs. Davenport is retired.

Davis, Cornelia

Cornelia "Connie" Davis was born in Tennessee and received her B.S. from the University of Illinois and an M.A. from Loyola University in Chicago, where she taught in the public schools and served as a vice principal. After she and her husband moved to California in the late 1950s, Mrs. Davis devoted herself to raising her children and to involvement with several charities. She had two sons and several grandchildren. The daughter she mentions in her letter became a physician. Mrs. Davis died in 2001.

Davis, George

George T. Davis was born in Montgomery, Alabama, in 1922. His family reports that he was a highly decorated World War II veteran who flew a B-24 bomber in over fifty missions. A high school English teacher, Mr. Davis had two children. He died of a heart attack in 1979.

Davis, Tony and Mrs. Donald K. Davis

Shortly after the letter was written, the Davis family moved to Idaho. Tony's mother was a great fan of the Kennedy family. Her son recalls that "Mom was patriotic and loving of her country and of her President." Tony Davis still lives in Idaho. His mother has since passed away.

Delaney, Henry H.

Henry Delaney enjoyed history and music. The daughter he mentioned in his letter died of cancer in 1982 after a successful career at Florida Atlantic University. Mr. Delaney died the same year.

Diamond, Ellen

Ellen has counseled disadvantaged youths in their career and education decisions, and worked as a peer counselor to help colleagues face and resolve personal issues. She has always lived in New York City. Now retired, she plays piano, sings with a chorus, and is a proud and loving aunt and great-aunt.

Diaz, Alden

Alden Diaz attended Reed College when he wrote to Mrs. Kennedy. He died in 1989.

DiGeorgio, Susan

Susan DiGeorgio earned advanced degrees in chemistry and pharmacy. She is the director of pharmacy at a mental health hospital. She also draws and paints and has shown her work at various locations in Connecticut.

Donnally, Mrs. Emma

Emma Donnally was born in August 1872 in West Virginia. She was married twice and died at the age of ninety-two in 1965 in Charleston, West Virginia. She is survived by several descendants.

Dryden, Jane

Jane Dryden's father was a physician who had done his internship at Parkland Hospital. She recalls imagining as a child that her father might have been able to save President Kennedy had he been on duty on November 22. Ms. Dryden graduated from college, married a physician, and has three children. She is involved in pastoral care work and lives in Texas.

Dudley, Grace

Grace Dudley was thirteen when she wrote to Mrs. Kennedy. Today she is a veterinary technician and lives in Connecticut with her husband, where they raise, train, and exhibit purebred dogs as a hobby.

Dumais, Roland

Roland Dumais was born in September 1919. Originally from Somersworth, New Hampshire, he served in the Army Air Corps as a fighter pilot during World War II. Mr. Dumais was shot down over Austria and spent six months in a prison camp. After the war, he was a corporate pilot. He was the primary campaign copilot on John F. Kennedy's private plane, *The Caroline*, during the 1960 Democratic presidential race, logging some 250,000 miles. He reported that he enjoyed the experience but also admitted that the pace was grueling. Mr. Dumais died in October 1972.

Emery, Harry T.

Harry T. Emery was born on February 23, 1915, one of five children, to a coal-mining engineer family and lived in Kentucky, Tennessee, and southern Illinois in coal-mining areas. He was a highly decorated Marine in World War II and worked for Field Packing Company in Chicago and later in the aircraft industry in California. He died in California at the age of seventy-five.

Emmitt, Tom

Tom Emmitt was a singer, composer, and conductor who played violin and viola, wrote criticism, and taught music. He changed careers in his forties when he started a family and became involved in mental health, spending the last twenty years of his professional life directing the Telecommunication Center for the California State Department of Health. He died in 1995.

Ernst, William and family

Fifteen months after composing the Ernst family letter to Mrs. Kennedy, William Ernst died suddenly of heart failure at the age of fifty-five, a condition, his son reports, "possibly induced by the rigors of his concentration camp incarceration." Mr. Ernst served as a GI in the Third Army under General George S. Patton Jr. His son recalls that "driving Nazi forces back across France and pursuing them into the heart of Germany gave him great satisfaction and a sense of retribution. Ironically, he was able to participate in the liberation of the very concentration camp in which he had been held prisoner." His wife, Pauline, died in 2005. His children, Katherine and Douglas, live in northern California.

Evers, Mrs. Medgar

Born in 1933 in Vicksburg, Mississippi, Myrlie Evers was a student at Alcorn A&M College when she met Medgar Evers. They were married in 1951 and moved to Jackson, Mississippi, in 1954. Mr. Evers served as state field secretary for the NAACP in Mississippi. Myrlie assisted him in his civil rights activities. They had three children. Mrs. Evers returned to college after her husband's death and went on to remarry and have a full professional life. She served as director of consumer affairs for the Atlantic Richfield Company, served on the Los Angeles Board of Public Works, and as chairman of the NAACP between 1995 and 1998. She twice ran unsuccessfully for Congress and is the author of two books. She was actively involved in efforts to retry Byron De La Beckwith, a Ku Klux Klansman and long the suspected assassin of her husband. Beckwith was convicted of the crime in 1994, thirty-one years after Medgar Evers's death.

Farquhar, Donald

Donald Farquhar was acting chief of industrial engineering at Electric Boat in Groton, Connecticut, and later acting director of industrial engineering at Sikorsky Aircraft in Stratford, Connecticut. He is the father of a son and a daughter. He coached softball and baseball in Ledyard/Groton, Connecticut, for thirty years. Today he remembers Kennedy's Presidency as a time when "all the bitterness and divisiveness that is prevalent today" was absent.

Fiola, Mr. and Mrs. Roland A.

The Fiolas had seven children, three of whom were born after 1963. Mrs. Fiola worked as a school bus driver in the Los Angeles area for nearly thirty years. She had rival gang members on her bus and gained their respect by learning each of their names and taking an interest in them as individuals. When her husband was killed on his job in 1989, several of these students rallied around her to provide support. Mrs. Fiola has eight grandchildren.

Floodas, Mr. and Mrs. John J.
Katherine Floodas remembers her husband as "a very deep person," who "enjoyed fishing, camping, and reading about the history of our country. He would have been very happy that someone appreciated something he wrote. He passed away on January 13, 1991."

Fowler, Dona
Dona Fowler's brother recollects: "My sister was always the most interesting person I knew. You always knew where you stood with Dona. If you were worried about the answer to your question, that it might be completely opposite of your point of view; don't ask my sister. She would always tell you exactly how she felt. The death of Robert Kennedy was devastating for my sister also."

Gambardelli, Arthur
Arthur Gambardelli was blind from birth. He and his wife had no children. His nephew recalls that Mr. Gambardelli took a great interest in the Kennedys. He died in 1979.

Garfield, Rudolph
Rudolph Garfield, born in 1920, is the great-grandson of James A. Garfield, the twentieth President of the United States. A graduate of Williams College, Mr. Garfield had a career in business and finance. He has been a trustee of several educational institutions and was a founding trustee of the Hattie Larlham Foundation, which provides assistance to children and adults with major developmental and physical disabilities. He is the father of three children and resides in Cleveland Heights, Ohio. He observes, "In reading my note after all these years my thoughts have not changed."

Gatewood, Chris K.
Christopher Gatewood was a furniture upholsterer who took an active interest in Democratic politics. He had two daughters and was the devoted grandfather of five. He died in 2003.

Geist, Mrs. Carroll A.
Widowed in 1971, Elizabeth Toland Geist was active in the PTA, with her church, and in the garden club. She had two sons, six grandchildren, and a great grandson. She died in 2001 and is remembered by her son as a "very positive person" who "tried to help all in need" and who was deeply devoted to her family.

Gidion, Gabriele
Gabriele Gidion lives in New York.

Givens, Ruby
Her daughter-in-law describes Mrs. Ruby Givens as a "a very prayerful mother" who raised ten children of her own as well as two other children in her community. Her second oldest daughter, Ruby, returned home safely from the U.S. Navy and went on to become a physical therapist. Her two oldest sons enlisted in the U.S. Army. One, who had previously served in Korea, was seriously wounded in Vietnam but recovered. Her family notes, "The actual biography of this caring woman and her children is an interesting book in itself." When Mrs. Ruby Givens passed away in 1992, she was still living in the home she mentioned in her letter to Mrs. Kennedy.

Glassner, Mary
Mary Glassner was born in June 1917, lived in Kansas City, and then in Denver. She had two daughters and a son. She provided day care to countless children through the years and was known as "Big Mom"—a mother and friend to the many for whom she cared. She died at the age of ninety-one in May 2009.

Glimpse, Nancy, Kenneth, Rick, Brandon, and Perry
Nancy Glimpse was, in her words, "a thirty-year-old stay-at-home mom with three boys—six, five, and four months" when she wrote to Mrs. Kennedy. She divorced a few years later and remarried. She spent twenty years in Yuma, New Mexico, working as a banker, and retired to Alpine, Texas, with her husband, who has since passed away. "During our time together," she recalls, "we water-skied, learned to fly, built a dune buggy named Myrtle, did some farming, lived full-time, and traveled in our travel trailer for several years, then spent a couple of years building a house." She now lives in New Mexico.

Golub, Shirley
Shirley Golub remembers seeing Jacqueline Kennedy Onassis walking on the Upper East Side of Manhattan and jogging in Central Park. "I'll never forget her dignified persona and intelligence—her love for the arts." She inspired Mrs. Golub's own involvement with her local arts council.

Gonzales, Henry
Henry Gonzales emigrated with his parents from Mexico to Texas when he was two years old. He became a stamp and coin collector at twelve. He worked as a projectionist in a movie house in Refugio, Texas, before serving in the U.S. Army during the Second World War. Henry later joined the U.S. Postal Service, where he worked until his retirement in 1976. He served as president of the El Paso Philatelic Society, traveled to many parts of the country, where he exhibited his stamps and first-day covers, and corresponded with other stamp collectors around the world.

He had four children who remember his gentleness, love of education, and devotion as a grandfather. His wife still lives in Texas.

Grumblatt, Douglas

In 1963, Doug Grumblatt recalls, he was "just starting the sixth grade in the city of Orange, California. . . . Before Dallas, my teacher was introducing poetry to the class. Many of us could relate to the works by Robert Frost, because we saw him on TV in the inauguration, and I think he had passed away earlier that year. In my condolence letter, I attempted a play on words from Mr. Frost's 'Stopping by Woods on a Snowy Evening.'" Mr. Grumblatt recalls writing an earlier letter to Senator John Kennedy during a 1960 campaign stop through Erie, Pennsylvania. On that occasion, he thanked the senator for visiting his town and giving him a campaign souvenir, which Doug continues to use today—a tie clasp in the shape of PT 109. Mr. Grumblatt is an engineer and lives in California.

Gum, Perry C.

Perry Gum was a retired farmer living in West Virginia when he wrote to Mrs. Kennedy. He had four children.

Hammonds, Mr. and Mrs. R.T.

Robert Taylor and Ruby Pearl Hammonds lived on a family farm in Texas where they had four children. Their grandson Dwayne recalls that when the Hammonds were informed of their son's death, they were presented with an American flag. Their grandson writes that after the terrorist attacks on September 11, "the lady in his life" found the flag in the attic and not knowing its provenance, hung it outside their home in Wyoming where it was shredded by the wind. At a ceremony organized to dispose of the flag respectfully, he writes, "almost one hundred cowboys" attended "and when we burned the flag there wasn't a dry eye at the place." Robert Hammonds died in 1979, ten years after his wife.

Hanrahan, Stephen J.

Stephen Hanrahan served in the U.S. Army during World War II. At the time of his letter to Mrs. Kennedy, he was serving time for grand larceny. He died in 1991.

Harayda, Eileen

Eileen Harayda was fourteen years old when she wrote to Mrs. Kennedy. She graduated from college in 1972 and joined the U.S. Navy. She served as an active duty nurse for twenty years and currently works as a nursing supervisor in a long-term-care facility. She is married and the mother of two children.

Harper, Mr. and Mrs. J.

The Harpers were both raised on farms around Gainesville, Texas. Their daughter notes that they "both had to quit school early to help pick cotton." Mr. Harper came to Teague, Texas, to work for the railroad and served as a conductor and brakeman on passenger trains. The couple had four children—all girls. Mrs. Harper died in 1985 at the age of eighty-five. She greatly admired JFK.

Harris, Charles A.
Charles Harris's daughter recalls that "it always remained a true honor for my dad to have served with John F. Kennedy during WWII on PT 109. He felt Kennedy had saved all their lives and was an American hero."

Harris, Mrs. Henry B. (Irene)
Irene Wallach Harris was born in 1876. Her husband, Henry, was a noted theatrical producer in New York City. They were both passengers on the *Titanic*. Her husband helped her into the last lifeboat lowered from the ocean liner and waved goodbye to her from the deck. She never saw him again. With initial success, Mrs. Harris took over her husband's theatrical business. She became one of the first women theatrical producers in New York and was extremely successful. However, she lost the theater she owned during the Great Depression and much of her wealth. She remarried several times. Mrs. Harris died in 1969.

Hebb, Jo
Betty Jo Hebb "enjoyed singing, traveling, and reading," according to her daughter. Her health improved and she lived almost seven years after she wrote to Mrs. Kennedy. She died of cancer in 1970 at the age of thirty-six.

Hemmerle, Patricia Anne
Patricia Hemmerle received a B.S. in chemistry from Indiana University and an M.B.A. from Butler University. She has worked for over thirty years in the field of health care. "In spring of 2008," she writes, "I had the privilege of voting for Hillary Clinton in the presidential primary. To my fifth-grade classmates—I told you so and that was the end of that."

Henrikson, Mike
Mike Henrikson writes: "As a high school dropout, with only ambition of being a farmer, my life was to change dramatically. The President's assassination was a historically world event; and the state funeral, of which I was honored to be a part, impacted my life and my perspective of life tremendously. My life realized a purpose, and I eventually graduated from college and was seriously involved in local politics. Although thankful for my changes, I will never forget that fateful, rainy, silent day, so long ago. While standing at the gravesite watching the funeral procession slowly moving across the Lincoln Memorial Bridge, I could hear only those distant drums echoing across the river . . . that sad march of a fallen hero. Now, as I think of all the extraordinary world events since 1963, I wonder of what could have been?"

Hesch, Mildred F.

Mildred Hesch is ninety-three years old and a widow. She raised three children of whom she is very proud. President Kennedy was the same age as her husband, and the Kennedy children were the same age as her children—a fact, she remembers, that made her "doubly sad" in November 1963. She still writes in a journal daily.

Hilliard, Jerrine

Mrs. Hilliard was a devout Christian who sang hymns while doing housework, opened her home to youth groups, to students from Hendrix College on "weekend mission trips," and at Christmas to children from the Methodist Orphanage. Her daughter remembers her mother's oft-stated conviction that "we are not rich in money, but we are rich in family, friends, and the love of God."

Hirsch, Janis

Janis Hirsch has had a very successful career as a sitcom writer. She lives in Los Angeles with her husband and son. She reports that she is active on behalf of a range of causes, from equal rights for the disabled to marriage equality. "I still sing 'You Gotta Have Heart,'" she writes, "when needed."

Hoerl, Norbert A.

Norbert Hoerl served in the military during World War II and Korea. He and his wife had four children, including the daughter he mentions in his letter. Mr. Hoerl played the piano and enjoyed drawing. He died in 1974.

Holman, Clinton Hale

His niece remembers Clinton Hale as a quiet man who was known as "Uncle Hickory." Although he was disabled, he made a living shining shoes and selling cigars in a downtown bus station. He lived with his extended family, never married, and had no children.

Holsey, Mrs. Magdalene

Magdalene Holsey lives in the South.

Housley, Mrs. Lennie Gore

Lennie Gore Housley Rogers was born in 1920. Although she had been married, she raised her three sons and a daughter mostly as a single parent, working very hard doing housekeeping for motels and in a sewing factory to support her family. Her family remembers that she never complained despite having to struggle to make ends meet. She revered President Kennedy and treasured a book she purchased about him. Her sons were all in the armed services, one son being called to be a guard at the inauguration parade route for President Johnson. Lennie treasured a trip she was able to make with her sister to see JFK's grave at Arlington National Cemetery. She died in July 1973 at the age of fifty-three.

Hughes, Langston

Langston Hughes was a renowned poet and writer. Born in 1902 in Joplin, Missouri, and raised in his early life by his grandmother, he moved to Lincoln, Illinois, to live with his mother when he was a teenager. The family later moved to Cleveland, Ohio. After high school, he lived in Mexico, attended Columbia University, worked at various odd jobs, and traveled. He published his first book of poetry in 1926 and later graduated from Lincoln University in Pennsylvania. He became a leading figure in the Harlem Renaissance. Langston Hughes died in 1967.

Jackson, Katherine Dowd

Mrs. Jackson grew up on a farm in North Carolina, married and had four children. Her oldest son died in the Korean War. Throughout her life, she was actively engaged in her church and her community. She taught Sunday school and was involved with the NAACP. She died in 1995 at the age of 87. One of her grandsons is currently in a graduate program at Harvard University.

Jacobs, Mrs. Howard A.

Mrs. Jacobs is still living in Illinois and works at the Lego store. She continues to pursue her passion for genealogy and enjoys being a grandmother to her four grandchildren. Two of her grandchildren are twins, and the others were born on the same day, five years apart.

Jakusik, John H.

John Jakusik was an avid fisherman who loved to read about history, politics, and religion. He was a family man devoted to his wife, Irene, and his two sons and grandchildren. After retiring from his work with the City of New Bedford, he continued to keep abreast of government policies and issues. He died in 1992 at the age of seventy-two.

Jones, Sandi

Sandi Jones recalls that she was a "young teen when I wrote this. Forty-six years later the day is still crystal clear. My interest in history continued to grow, nurtured by my father's passion for ephemera. Unfortunately, both the Kennedy spruces died, and just last year the Lincoln spruce succumbed to a fatal lightning bolt."

Junker, Ellen

Ellen Junker, who is a nurse, writes: "When I reread my letter, I remember a crushing sense that goodness has many enemies; that the very light, hope, poetry, and sense I saw in Jack Kennedy made him vulnerable. It was not a great feeling for a twelve-year-old, who was still in the good-versus-evil, all-versus-nothing, win-or-lose stage. I also wonder if Kennedy's assassination contributed to my passion for working in the field of death and dying."

Katzberg, Gloria
Gloria Katzberg is now eighty-three years old and worked full time until she was eighty. She enjoys writing. She made a scrapbook for her daughter about the assassination. "Rereading my letter," she commented, "brought tears to my eyes remembering the sadness felt by all."

Keat, Betty
Betty Keat was born in 1932 and died in 1996.

Keegin, Arch C.
Archibald C. Keegin began his career in the civil service system at the age of fourteen, when he was forced to leave school to help support his six siblings after his father died during the 1918 flu epidemic. He completed his schooling at night classes and went on to hold various clerical positions within the FBI and the Department of Justice, including personal secretary to J. Edgar Hoover. He retired as chief of the Division of Supplies and Printing, after forty years of service in the Department of Justice. His daughter observes that it was "no small wonder that Arch was dismayed that the singing by the many processing mourners at JFK's funeral had not been reported." He enjoyed singing and was considered a fine entertainer by his family, friends, and fellow workers.

Keller, Rev. David
After leaving Shageluk in 1968, David Keller worked with Alaska Native leaders advocating the federal Alaska Native Claims Settlement Act of 1991. He lives now in Weaverville, North Carolina, leads retreats, and is the author of a book on the power of personal prayer.

Kendler, Karen S.
Dr. Karen S. Camara, formerly Karen S. Kendler, was a Peace Corps volunteer in Nigeria. She served as a program director for low-income science students and later as a computer developer. She is the author of a book on coming-of-age in New York during the 1950s.

Kilmurry, Ellen
Ellen Kilmurry worked in public accounting for nearly twenty years before leaving corporate work to become involved in assisting homeless services agencies. She is currently the executive director of Southwest Chicago PADS, a nonprofit that serves nearly twenty-one hundred persons annually.

Klemkosky, Brenda
Since 1963, Brenda Klemkosky has married and had two sons and two granddaughters. She and her husband owned and operated a construction company in

Michigan. She lost her husband in 2001, and her sister was widowed a year later. She now lives with her sister and their two dogs.

Koop, Mrs. Patricia
Patricia Koop was from the Chicago area and moved to the state of Washington with her husband. They had two sons and were later divorced.

Lane, Suzan Elizabeth
Suzan Lane moved to Durango, Colorado, where she raised her three sons. She is "grateful for the life I get to live" and writes that "for Christmas, 1993, my mother surprised me with the gift of a rocker like President Kennedy's."

Legg, Jim
Jim Legg graduated from Washington and Lee University, where he majored in physics and math. He obtained an M.A. in physics and has worked for several aerospace companies in conjunction with NASA, including many spacecraft projects.

Lewis, Lillie M.
Lillie Mae Lewis was born in Georgia in May 1917. She was an agent for Central Life Insurance Company in Florida and a member of the Hopewell Baptist Church in Pompano Beach. She married three times and had one son. She died in June 1974 in Florida.

Link, Leonard L., Jr.
Leonard Link was single and had been driving for the United Parcel Service for less than one month when President Kennedy was assassinated. He married in 1968 and has three children and seven grandchildren. He recalls JFK with admiration to this day and mentions that he still misses him.

Lockeby, Janis
Janis Lockeby grew up in Philadelphia and in 1966 moved to California, where she met her future husband. Janis worked for American Airlines and retired after thirty years as a travel consultant. She was involved for many years in the dog rescue of golden retrievers. She and her husband of forty-three years live in California but also spend time in Idaho. She reports, "I am still a political 'wonk'!"

Lodge, George C.
George Lodge reports that "after serving in the Eisenhower and Kennedy administrations as assistant secretary of labor for international affairs and running as the Republican candidate for the U.S. Senate in Massachusetts in 1962, I became a professor at the Harvard Business School." He retired in 1997.

Longeneker, Grace
Grace Longeneker lives on the West Coast.

Louk, R. R.

Mr. Louk never attended church until after the death of his wife. His family notes that he then "never missed a service. . . . He mourned her death until his own in January 1972."

Lounsbery, Mrs. Anna and Anna Lounsbery

Ann Lounsbery Owens lives in Seattle and vividly recalls "being sent off from the Rose Garden by President Kennedy himself" in one of the first group of Peace Corps volunteers. She remembers that the group "truly felt ourselves to be goodwill ambassadors from America." She served in Makelle, Ethiopia. Her mother is no longer living.

Lowrey, Irene

Irene Lowrey was a wife and mother as well as a schoolteacher in La Porte, Texas, when she wrote to Mrs. Kennedy. She remained in La Porte until her death and had deep bonds with her family and community.

Lundstrom, Gretchen

After graduating from college, Gretchen Lundstrom had a successful career in teaching and librarianship, mostly at the University of Wisconsin in Madison. She has been active in politics. She has one daughter, two sons, and seven grandchildren. She retired in 2001, enjoys family history, and is active in church, cultural, and community programs.

Lynch, Sheila

Sheila J. Lynch was born on January 3, 1927, in Buffalo, New York, to George and Grace Lynch. She and her parents moved to Southern California. Sheila never married and died on December 5, 1982, in Van Nuys, California.

MacArthur, General Douglas

General MacArthur died less than a year after President Kennedy. Colonel William J. Davis, executive director of the General Douglas MacArthur Foundation, writes that General MacArthur "and President Kennedy served our country with 'Honor and Devotion' and must be remembered to future generations."

Macho, Mary and Adolph Sr.

Mr. Macho was a talented glassblower. He died in 1976; his wife died five years later.

Mackey, Vivian

Vivian Mackey was born on March 30, 1906, and was a teacher in the Seattle School District for many years. She had an avid interest in her family history, was an active member in genealogical societies, and actively took classes in genealogy until she was well into her eighties. She died in February 1994.

Manfre, Mrs. Pati and Vivian

Mrs. Manfre's granddaughter Vivian notes that "my grandmother is no longer with us but her foresight and determination moved her from a rural Italian village to the streets of New York. Her persistence gave me the opportunity to embrace the 'American Dream' and to appreciate this great nation I live in. We were taught to adhere to the words of President Jack Kennedy—'Ask not what your country can do for you, but what you can do for your country.'"

Martin, E. G.

Elwood G. Martin's son, Kenton, observes that his father "was a simple, ordinary person whose life and attitudes were forged by two extraordinary historical events: the Great Depression and World War II. During the Depression, he owned a small jewelry business in the tiny town of Cypress, Illinois, the town of his birth. The Depression, however, killed his business as it killed so many other dreams. Oddly enough, he saw World War II as a way to escape his economic plight in Illinois." Mr. Martin served in the Army Air Corps during the war and attempted to take up his career as a jeweler again. But his son notes, "The postwar recession put an end to that. Eventually, Dad was able to get by as a bus driver. Obviously John F. Kennedy and Dad were from two different worlds, but they were parallel worlds. Kennedy lost a brother in the war and faced its horrors firsthand, and although the Kennedy family did not suffer economically during the Depression, they had empathy for those who did. Thus Dad sensed a kindred spirit. My father was a Republican, but he admired the Kennedy family's nobility and courage. Kennedy was rich, but his social consciousness seemed to transcend class lines."

Matteson, Austin

Austin Matteson lives in the Pacific Northwest. Of President Kennedy, he writes: "He was a great one, as was his wife!"

McClain, David Blair

David McClain left high school in 1971 and earned a GED a short time later. He served in the navy and, he notes, "unloaded Marines to fight the Viet Cong. Got out in 1975 and went to work for Floyd Cleveland, owner of Home Appliance, and repaired and sold vacuum cleaners until 2008." He is the father of a daughter and tries "to enjoy life and my family."

McCormack, Mrs. Melbro R.

Melbro Ruth McCormack was born in the country of Jamaica in 1881 and coincidentally lived in Jamaica, New York, for a time. She died in April 1967.

McIver, Renee

Renee McIver lives in Washington State and is a widow with a grown daughter. She is retired but says, "Travel, reading, and the study of Spanish are still priorities."

Of 1963, she observes that "time, cynicism, too much reality have not dimmed that moment in time that helped propel me to write to Mrs. Kennedy."

McKenney, Daisy H.

Daisy McKenney was born in November 1914. She was a resident of Hohen Solms for over eighty years. She was named Woman of the Year by the St. Philip Baptist Church. Married and widowed twice, she died in December 2008.

McLean, Margaret

Margaret McLean was a housewife prior to her husband's death. She attended community college after she was widowed, took secretarial courses, and completed her education in record time. She remarried in 1973. She worked for an ink company and managed to send all her children to college. Retired at the age of seventy-two, she has now moved to an assisted-living facility.

McManus, Mrs. Dorothy

Dorothy McManus was an accomplished businesswoman, artist, sculptor, and novelist who lived to the age of ninety. Her son wrote a book about her life after her death, *Rancho El Contento*.

McMillen, Mary

Mary McMillen was eleven when she wrote to Mrs. Kennedy. She has been married for twenty-nine years and with her husband enjoys "church, our dogs, and camping."

McMurtry, Gertrude

Gertrude McMurtry was born in December 1897 in Greenbrier, Robertson County, Tennessee. She married Lanes McMurtry and lived in Cincinnati, Ohio, and Louisville and Dayton, Kentucky, during their marriage. She died on October 25, 1966, at the age of sixty-eight.

McNeil, Grady

Grady McNeil was born in June 1916 in Waycross, Georgia. He was one of nine children. He lived in Florida during his childhood and enlisted in the army in 1940. A World War II veteran, he lived in New York City. He died in September 1987.

Meader, Vaughn

Vaughn Meader's wife comments: "After Kennedy's death people would still ask my husband to imitate the late President. Vaughn (family and friends knew him as Abbott), who admired and was respectful of Mr. Kennedy, always answered, 'Sorry, the man hasn't said anything lately.'" Mr. Meader died in 2004.

Melder, Mrs. M. F. and Ray Melder

Ray Melder became a management consultant after serving in the army. He was married in Germany in 1964 and recently celebrated his forty-fifth wedding anniversary. His mother was a portrait artist. She died in 1996.

Mesaros, Margaret
Margaret Mesaros remained a widow for seventeen years. She remarried and was widowed once again in 1986 after a six-year marriage to a childhood friend. Margaret enjoyed her children and two grandchildren in her remaining years. She died in May of 1996.

Metzger, Regina and Louis
Regina Metzger loved to travel. Her daughter Gloria recalls many memorable trips with her mother, visiting Africa, Europe, and Asia. Mrs. Metzger was widowed in 1965. She passed away in 1980.

Michalski, Lorraine J., and Thomas A. Michalski
Lorraine Michalski was married in 1968 and has two children and four grandchildren. She lives in Huntington, Long Island, and enjoys cooking, knitting, computers, traveling, psychology, and playing with her grandchildren. Her brother Thomas observes that "America needs memories now" of the early 1960s. He continues: "A young family suffering so much, but the wife showing the whole nation that there is always hope and anything is possible when we all work together."

Milano, Helen M.
Helen Milano never remarried. She had eight grandchildren, four boys and four girls, who gave her "great joy," according to her family. Her family reports that "she continued to pray for Mrs. Kennedy and her family, never forgetting the losses they both experienced."

Mitchell, Eileen
R. Eileen Mitchell graduated from North Texas State University in 1966 with a major in English. She taught English and now works as a paralegal in New Orleans, where she also teaches paralegal studies. She remembers that day in 1963 as if she were still standing in the window of her dorm room. "It was a pivotal point in our history," she comments. "It never occurred to me that life was that dangerous and that our President could be assassinated."

"Mrs. American Citizen"
This anonymous letter was postmarked "Houston," but the writer provided no signature or return address.

Nash, Mrs. Frances
Mrs. Nash has retired from the U.S. Postal Service and lives in Michigan.

"A Negro Who beleave In God"
This letter was sent anonymously with no signature, date, or return address.

Nichols, Linda Gayle and Donnie
Linda Nichols has been married thirty-eight years. She has a son and a daughter and five granddaughters. She was a homemaker until 2001 when she began working for a small oil and gas company. Her brother Donald works for Union Pacific Railroad and has a son.

Nichols, Ruby K.
Ruby Nichols is ninety-two years old and lives in Illinois.

Nies, Mary F.
Mary Ford Nies was born in May 1913 in Rockingham, North Carolina. She moved to New York City with her family and graduated from Hunter College. She joined the staff of *Natural History* magazine, where she met her future husband Frederick Nies, a freelance writer. They married in 1940. Mary became the editor of a monthly magazine published by the Pepsi-Cola Company in New York. After the couple moved to Chapel Hill, North Carolina, she edited a quarterly magazine. She was director of public relations for the North Carolina and later South Carolina Heart Association. Active in the civil rights movement and the local humane society, she also worked in antipoverty endeavors with her second husband, whom she married after she was widowed from Mr. Nies. She died in January 2007 in Greensboro, North Carolina.

Oakey, Carol
Carol Oakey was twenty years old and was a student at North Texas State when she wrote to Mrs. Kennedy. She was a student-teacher in a suburb of Dallas, and with a friend managed to get November 22 off so that she could go downtown and see President Kennedy. She remembers the day "just like it was yesterday." She heard the news of the President's assassination over a police radio as she waited at the Trade Mart and saw the limousine speeding toward Parkland Hospital. The whole experience, she said, turned her into a Democrat. She joined the Peace Corps after college and served in Nigeria. Carol later taught English and history in the United States. In 1974 she began working as a news dictationist at the *Washington Post*. She worked there until 2003. She has two grown children and three grandchildren.

O'Connor, Peter N.
Peter O'Connor survived his tour of duty in Vietnam in 1964. He died of cancer in 1978, shortly before his forty-fourth birthday. His widow recalls that "as a native of Boston, he was especially proud of President Kennedy."

Oglesby, Susie

Susie Hill Oglesby was born in March 1893 to George and Eliza Jane Hill in Athens, Georgia, one of seven children. She married, moved to Tennessee in 1922, and was widowed at a young age, supporting herself as a domestic worker and maid. She died in Chattanooga, Tennessee, in April 1971.

Oldham, Mrs. Helen

Helen Oldham and her husband had seven children and twenty-one grandchildren. She loved watching baseball and was, her grandson reports, "a huge fan of the New York Yankees." She lived in the Shiloh community in Montgomery County near Clarksville, Tennessee. She was, her grandson notes, a "yellow dog Democrat." She still lived in the same house when she died in November 1972 that she inhabited when she wrote to Mrs. Kennedy. "She had a picture of President Kennedy above her bed until she died," her grandson writes. "She had cut it out of the local newspaper, the *Leaf Chronicle*, when he was elected President."

Ostby, Janeen

Janeen Ostby and her husband adopted three children and now have two grandchildren. Janeen earned a B.A. in social work in 1980. She retired as a probation officer in Orange County, California, where she explores the arts, paints watercolors, and is currently writing a book about her aunt who was an artist.

Palumbo, Jo-Ann

Jo-Ann Palumbo and her husband own a construction company. They have been married for forty years and have two children and a grandson. Jo-Ann is also an artist who works in oil painting.

Parrish, Melvin

Melvin Parrish was an elementary school teacher for many years in Fancy Prairie, Illinois, and enjoyed history, photography, and gardening.

Parsons, Mrs. Catherine

Catherine Parsons grew up in Shelby, North Carolina, in a mixed-race family. Her parents were white but adopted a Native American child and had an African American foster daughter. Mrs. Parsons attended Winthrop College and earned a degree in social work and journalism. She also held degrees from the University of North Carolina. She taught high school English and history, was very active in community and charitable affairs, and was a staunch supporter of civil rights, especially during the 1960s Greensboro sit-in, in which she was a participant. She died in August 2005, leaving two sons and a large and devoted extended family.

Paul, Mrs. Jerome
Rose Lee Murphy Paul was born in Cincinnati, Ohio, in 1933 and grew up in Bates-ville, Indiana. Rosie enjoyed foster parenting and had five children of her own. She was diagnosed with breast cancer and lived with it for ten years. She passed away in July 1998 in Batesville, Indiana.

Pinkney, Grace
Grace Pinkney is remembered by her family as "well liked by everyone." She and her husband, George Pinkney, raised two children, Peggy and George, in Harrisburg, Pennsylvania. She dearly loved her grandchildren but did not live to see them grow up. She died shortly after writing the letter to Mrs. Kennedy. Portraits of President Kennedy and Martin Luther King hung on the wall of her daughter's home for many years.

Poberezny, Doris
Doris Poberezny moved to Florida in 1965. She became a cosmetologist, motel owner, and a realtor, and had a property management company and a tavern. She moved to Canada, and then back to the United States, where she worked for a large brokerage firm in Detroit. The firm went under, she reports, "on the horrifying day, September 11, 2001." She was then seventy-three years old.

Pond, Kate
Kate Pond graduated from Cornell in 1978. She earned an M.A. in journalism at the University of Michigan. She spent, she writes, "some years on the crew in the film industry and right now I raise olives trees and cure their fruit in Southern California."

Pucka, Elizabeth
Elizabeth Pucka died in 1992.

Quan, Mary
Mary Ng Quan's family describes her as "elfin, magical—like no one anyone has ever met." They comment that Mary "was the third of ten Chinese brothers and sis-ters who barely survived the devastating Battle of Manila in the Philippines in 1945 and the first of her family to emigrate to the United States. She embraced this land with the fervor and moral certitude of a patriot. This was due to her upbringing as the product of an American school in Manila established by the Thomasites. It was the only school in the Philippines that accepted not just American kids but also children of other nationalities, a mini-America, if you like. Mary Ng Quan loved life's little treasures—pausing in daily walks to church to admire the tiny flower on a weed sprouting through a crack in a sidewalk and pausing to pick up a dis-carded key chain or chat with a cat. She was the master of the unexpected—the wry

remark that cut to the truth and that left her listener howling in delight. She was an editor at McGraw-Hill for many decades, an active volunteer at her local Catholic church, and a beloved sister and aunt. Her brothers and sisters and over twenty nieces and nephews looked forward to her quarterly cards and letters filled with wit, truth, advice, and love. Mary was a good and caring woman who loved her adopted country and took inspiration from both the life and death of her President, John F. Kennedy. In 1991 she moved from New York City to be with family in Chicago and died there in 1995."

Radell, Kevin
Kevin Radell moved to New York City after doing graduate work at the London School of Economics and Northwestern School of Business. He is an investment banker and a fine art adviser.

Ralston, Frances
Frances J. Ralston was born in February 1911 in Tennessee and was one of twelve children. She married Robert E. Ralston. They had a son and a daughter. Mrs. Ralston died in March 1977 in Middlesboro, Kentucky, at the age of sixty-six.

Rassche, Mrs. Catherine
Catherine Rassche's nephew remembers his aunt as a "good and caring woman who loved her adopted country and took inspiration from both the life and death of her President, John F. Kennedy. In 1991 she moved from New York City to be with family in Chicago and died there in 1995."

Ray, Mrs. Whitley, and Larry Jackson
Mrs. Whitley Ray had two sons and a daughter. She was widowed twice. Her son Larry Jackson is now retired as a court administrator and Naval Reserve Commander. His mother died from complications due to breast cancer in 1990.

Rice, Leonard C.
Leonard Rice was born in Nucla, Colorado, in 1918 and moved to California when he was eight years old, where he attended school. He served in the U.S. Navy for twenty years and was discharged as chief boatswain's mate in 1967. He then became a letter carrier with the Porterville Post Office and retired in 1978. He was the father of seven children, and an avid golfer who enjoyed researching his family history and playing pinochle. His widow notes that "after being sent home from the hospital without any hope, he wrote his own obituary." He died in 1995.

Rimer, Barbara
Barbara Rimer is the Dean of the University of North Carolina's Gillings School of Public Health and Alumni Distinguished Professor. She is a leading scholar, teacher, administrator, and consultant in the field of public health and notes that

"John F. Kennedy's vision of service to country" has guided her "throughout her public health career."

Robinson, Ethel M.

Ethel M. Robinson, her family writes, "was born December 7, 1910, in Phoenixville, Pennsylvania, and died in June 1978. She was artistic, wrote poetry, and worked for Western Electric. She and her husband, Samuel, raised two children. She was an avid antique collector and always willing to help anyone in need. Ethel was a strong advocate of education and cultural pursuits, and believed in being self-reliant. She was way ahead of her time in terms of being a strong, independent, hard-working woman. She loved her family and her country, and was very patriotic. The Kennedys were a big focal point of her life and highly revered in her family. When JFK made a campaign stop in Norristown, Ethel, Samuel, and daughter Joan Tolliver all were pleased to have shaken hands with him. She was much loved and is missed by her family." Her granddaughter adds that "Mrs. Kennedy was such an outstanding role model for me, the women in my family, and our nation. . . . My grandmother appreciated Jacqueline Kennedy."

Robinson, Mr. and Mrs. Hugh B., Jr.

Hugh Robinson was born and raised in Philadelphia. He was a high school track star and played the violin. He served in the U.S. Navy and was a veteran of the Korean War. A cook by trade, he worked in a Veteran's Hospital in Pennsylvania. He died in 2001. His wife, Rosa, died three years later. The Robinsons are survived by their son, Bruce.

Rock, Leonard F.

Leonard Rock, who was a victim of polio, ran a magazine subscription service. He died in 1978.

Rosenberg, Martin

After graduating from the University of Massachusetts–Amherst in 1965, Martin J. Rosenberg went on to earn his Ph.D. in biology at SUNY–Stony Brook. For the next thirty-five years, he served on the faculty of the department of biology at Case Western Reserve University in Cleveland, Ohio. His primary academic interests during this period were herpetology and human anatomy/physiology. Following retirement in 2006, Dr. Rosenberg has worked part-time doing nonmedical care and companionship for the homebound elderly and those in assisted-living facilities. As with many others, he still remembers exactly where he was and what he was doing on November 22, 1963.

Ross, Martha

Martha Ross was born on May 10, 1890, in Georgia. She was a sharecropper and moved north during the 1950s. She married Jessie Ross and settled in New Haven,

Connecticut. Her great-grandson remembers her vividly as a strong and determined woman who was truly a "matriarch" to her family. She taught him to read by focusing on the Bible. Martha Ross died on August 6, 1981, in New Haven, at the age of ninety-two, and is survived by her large extended family.

Runnals, Thomas

Thomas Runnals was born of immigrant parents from England. He was raised during the Depression and left college during World War II to join the service. He was able to visit his relatives again when stationed in England. He participated in the Normandy invasion. After the war, Mr. Runnals returned to college on the GI Bill and became a high school chemistry teacher.

Russell, J. E. Y.

J. E. Y. Russell was born in 1888 and had nine siblings. His father was a traveling preacher in Texas. He toured with his father and would sing. He settled down with his wife in Dallas. When his father retired from preaching, they ran a filling station together and a number of other businesses. They raised hogs at one point. Mr. Russell lost everything in the Depression and did whatever he could to earn money and keep his family together. He lived until his late eighties.

Sanders, Donald

Donald Sanders served in Company F of the 43rd Infantry during World War II and fought for three years in the islands of the Pacific, completing his tour of duty when the war ended in 1945. Mr. Sanders and his wife of thirty-three years—Rosanne Bryce Sanders—are both retired from Eastern Illinois University in Charleston. Mr. Sanders was the head concrete finisher on the grounds, and his wife was a professor in the College of Business. He reports that "he participates regularly in parades honoring WWII veterans and frequently gives talks to students (of all ages) on his experiences in the Pacific area of the War."

Schechter, Jean

Jean Schechter; her husband, Philip; and her small daughter, Barbara, survived the Nazi occupation of Poland. She is featured in the book *Heroes of the Holocaust*. Her daughter reports that Mrs. Schechter never lost her affection for President Kennedy.

Schwen, Marcia

Marcia Schwen was a nineteen-year-old student at Gustavus Adolphus College when she studied in Mexico on a Carnegie Foundation grant and a first-year teacher at Long Island Lutheran High School in New York when President Kennedy was assassinated. She has served as an award-winning editor of several community newspapers in New York and Rhode Island, as well as an editor of Amnesty International USA publications. A mother and grandmother, she lives in Snug Harbor, Rhode Island. She is still close to the family from Guadalajara she mentioned in the letter.

Scott, Mrs. E.

Violet Phinisse Scott was born on May 22, 1922, and lived in Mississippi, Iowa, Michigan, and Washington. Her daughter describes her as a "loving wife and mother who was a positive thinker, a wonderful hostess, and a great cook, and someone who loved her country." She made "anybody smile and feel joy and peace." Her husband, Elisha Scott Jr., was an African American attorney in Flint, Michigan, the son and brother of the attorneys in Topeka, Kansas, who played a major role in the *Brown v. Board of Education* landmark lawsuit. She was married twice, to Carl Ross and Elisha Scott Jr., by whom she had three children, Michael Ross, Tonya, and Elisha Roy Scott. She died on February 28, 1981, in Seattle at the age of fifty-eight.

Seiler, Ira

Ira Seiler was a second-year pediatric resident at Georgetown University Hospital on Thanksgiving Day in 1960 when Jacqueline Kennedy was admitted for an emergency caesarian section. When John F. Kennedy Jr. was born, he did not breathe on his own immediately and was intubated by Dr. Seiler. He recalls vividly breathing "air into the lungs of the baby." JFK Jr. was transferred to the intensive care unit where Dr. Seiler cared for him until his discharge two weeks later. Shortly thereafter Dr. Seiler received a handwritten thank-you note from President-elect Kennedy and in January a registered letter with tickets to various inaugural events, sent at the request of JFK. On January 13, 1961, a special-delivery letter arrived with instructions on how to collect tickets to the inauguration for him and his wife. On the day Kennedy took the oath of office, the Seilers sat on the platform "behind Mrs. Roosevelt and next to Adlai Stevenson." A few weeks later as he visited a newborn at Georgetown Hospital, Dr. Seiler "mentioned to one of the nurses how impressed I was in receiving all the inaugural invitations. She informed me that she had written to the Inaugural Committee that if I had not been there the baby would have died." Dr. Seiler is retired from medical practice after a long career and lives in Florida.

Sheldon, Judy

Judy Sheldon's sister Anita recalls that "Judy had a heart of gold, and she was at her best when someone she loved needed comfort. After 1963 she continued to work as an office manager for an important accounting firm and she also continued to study voice, which she had done as a hobby for many years. Opera was her love and she performed several times, in the fifties and early sixties, with other students, first at the Music School of the Henry Street Settlement and then at the Mannes School of Music, both in New York. As the years went by she gave up her music studies, but she continued to sing at family occasions and friends' parties. She and I were planning to retire when she reached sixty-five, and she was looking forward to spending the rest of her life in our little house in East Hampton. Unfortunately she never made it: cancer took her in 1999. I am sure she is in a better place now, but I will never stop missing her."

Shelmire, Dr. and Mrs. David

Dr. Shelmire was born in Texas and attended Highland Park High School and the University of Texas. He came from a family of doctors and earned his medical degree in 1960 at Washington University Medical School. He completed his internship at Baylor and his residency at the University of Michigan Hospital. Dr. Shelmire is now seventy-five years old and still practicing dermatology full time.

Sherman, Estelle

Born into a Polish immigrant and coal-mining family in Forestville, Pennsylvania, Estelle Sherman lived most of her life with her husband, George, in Spring Lake Heights, New Jersey. She raised two daughters and a son, and was a devoted and much-beloved figure in her family and community. Her daughter recalls that her mother "visited the elderly in nursing homes, hand-decorated Polish Easter eggs for family and neighbors, and always made candied apples for the neighborhood children at Halloween. Some of those children returned years later with their own children to Trick or Treat for Mrs. Sherman's candied apples." Mrs. Sherman was also renowned for her letter writing. Her daughter writes that her mother "never forgot someone's birthday or an accomplishment in someone's life. Her family was fond of saying that she helped keep the post office in business." Mrs. Sherman died in October 2002.

Siegel, Natalie

Natalie Siegel lives in Florida. She enjoys playing bingo and spending time with friends.

Silverstein, Irving

Irving Silverstein was born in Brooklyn in 1913 and was the son of Russian Jews. He served in the army during World War II and was wounded in the Battle of the Bulge. He and his wife settled on Long Island, where they raised their two daughters. In addition to a sales career in the photographic industry, Mr. Silverstein loved to read, had a passion for golf, and was a lifelong Democrat active in politics. He died on November 25, 1985.

Simrin, Arlene

Mrs. Simrin is now elderly. Her daughter Stacey lives in New Jersey.

Skeats, Claudine R.

Claudine Rogers Skeats was born in November 1918. She married 1st Lieutenant Arthur E. Skeats Jr., a B-29 pilot who was killed in a plane crash on July 24, 1945. Their son, Arthur, was born after his father's death. As Secretary to the Base Commander at Brooks Air Force Base, Mrs. Skeats came to know all seven Mercury astronauts who came through the School of Aerospace Medicine at Brooks. She also knew LBJ. Claudine Skeats died in Texas on December 6, 2001, at the age of eighty-three.

Smith, Ernan H.

Ernan Smith continued to work at the power plant for the Latter Day Saints Hospital. He died in his seventies.

Smith, Mrs. Paul

Pat Smith lives in Pennsylvania. Since November 1963, she has had two more children, but also lost two children as well as a grandchild, and in 1998 her husband. She works part-time and during the week takes care of her eight-year-old grandchild. She reports that despite the losses she has experienced, she has had many positive times and is very proud of her two sons, one of whom served in Vietnam and the other in the Air Force. She comments that the Kennedy family gave her a sense of strength during difficult times.

Smith, Tommy

Tommy Smith and his brother played hooky on November 22, 1963, to go downtown to Dallas to see President Kennedy. He has kept a box of memorabilia that he collected at the time of the assassination—an event he still remembers vividly. He writes: "Through my teens, I was very interested in current events. School often took a backseat as I intensely followed the news, like standing along the Kennedy motorcade route or attending a session of the Jack Ruby murder trial. The dramatic, chaotic news events of the 1960s nurtured my dreams of a career in journalism. However, my first year in college included a class in geology, and I realized that my real dream was what it had always been as a boy, and that was to become a scientist. My love for current events and my passion for earth science are with me now, as I continue in a career involving environmental consulting and teaching at the college level. My commitment to keeping abreast of the latest scientific developments and passing those on to eager students is rooted in those tumultuous days of my teens, when calm and chaos became routine, when the cheers of the crowd and three rifle shots became one sound, and a boy watched as the world revealed itself in glory and shame."

Snell, Alma

Alma Snell was born in 1923. She farmed on the Fort Belknap Reservation with her husband, authored two books on the Crow way of life, and was an adviser on the creation of the National Museum of the American Indian. She died in 2008.

Snider, Mrs. Merlene

Merlene Snider was one of the first young women in her area of West Virginia to graduate from high school. She loved writing and music, and often penned short stories and songs. She suffered from Huntington's disease and eventually succumbed to it.

Sooby, Donna
Donna Sooby's family reports that she "enjoyed life and lived it to its fullest. She loved the outdoors: camping, boating, hiking, snowmobiling. She went on to earn a Ph.D. and to conduct research on leukemia. Ironically, she died from the disease in 1975 at the age of forty-three."

South, Mary
Mary South Certa grew up in Santa Clara, California, earned a bachelor's degree in English and a master's in education, and became a teacher. Her love of language and history, which began in the 1960s, continues today. Married for thirty-eight years and a mother of three, she lives in Campbell, California.

Spector, Pauline and Sol K.
Mrs. Spector's family writes: "Whenever a friend or family member had a loss, Pauline would write a letter expressing her condolences. She and Sol felt that the Kennedys were part of our family, and she felt moved to send the note to Mrs. Kennedy. Both Sol and Pauline have died but lived to take pleasure in their children and grandchildren."

Stafford, Mrs. Ruth
Mrs. Stafford's family notes that it "just broke her heart" when President Kennedy was assassinated. She is remembered with "great love" as a woman who "enjoyed life, especially her five grandchildren." She died on September 23, 1984, at the age of seventy-eight.

Stamos, Katherine and Spiro P.
Katherine and Spiro Stamos have been married for fifty-four years and currently reside in San Francisco. Spiro Stamos was a Hollywood studio violinist for over thirty years. In November 1963, the Los Angeles Chamber Orchestra was on a concert tour of Europe, sponsored by the U.S. State Department under the auspices of the Kennedy administration's fostering the arts and culture program. The "Jackie" referred to in the letter was Marilyn Horne, world-famous mezzo-soprano, and "Henry" was Henry Lewis, the African American conductor to whom she was married at that time.

Stanley-Brown, Katharine
Katharine Stanley-Brown was born in April 1892 and became an artist, musician, and writer, who, along with her husband, Rudolph Stanley-Brown, and two children lived in Washington from 1934 to 1966. Her husband, an architect, was

recruited from Cleveland by Franklin Delano Roosevelt to design public buildings throughout the country with "federal character," using indigenous materials. After his death in 1944, Katharine Stanley-Brown moved to New York and worked at Harper & Bros., who had published the books she and her husband had written and illustrated in the 1930s. She died in April 1972.

Starr, Morris
Morris Starr still vividly recalls the day of the Kennedy assassination. He is eighty-one years old, a businessman, and is still working.

Steinhart, John
John Shannon Steinhart was born in 1929 and earned a B.A. in economics from Harvard, where he was a varsity swimmer. He served in the U.S. Navy for four years and then studied geophysics at the University of Wisconsin–Madison, where he earned his Ph.D. in 1960. He was a scientist at the Carnegie Institution's Department of Terrestrial Magnetism in Washington, D.C., and served on the staff of the White House science adviser in the Johnson and Nixon administrations. He had a distinguished career at the University of Wisconsin from 1969 to his retirement in 1991. He died in 2003.

Stone, Helen
Helen Stone was of Dutch ancestry and married a Cherokee Indian. She worked in a laundry in Columbia, Missouri, and later as a cook at the University of Missouri. She was a wonderful storyteller and extremely dedicated to her seven children. Helen Stone died in 1990 and is greatly missed by those who knew and loved her. Mrs. Stone's daughter writes, "I often wished I could have her beautiful heart. She loved her family with all her heart, her country, and she loved John F. Kennedy, as we all did. She was a person who loved and had a lot of love to give."

Storll, Mrs. F.
Sarah S. Storll was born in 1894. She enjoyed reading, bingo, and card playing, often hosting card games at her home for her many close friends. She had one daughter, two grandchildren, and one great-grandchild. Her granddaughter remembers that Mrs. Storll "made history come alive to me from her recollections of the great events that occurred during her lifetime, such as the Depression, the *Titanic*, the *Lusitania*, Triangle Fire, and both world wars." She died in 1984.

Swain, Marzell
In 1924, at the age of two, Marzell Swain moved to Newark, New Jersey, with her parents. She graduated from South Side High School in 1940. A wife and mother, she enjoyed singing, reading, and the cinema. Her family notes that "Marzell was often called upon by family and friends to read or compose correspondence for them. The richness of her sentiment more than compensated for her lack of money,

especially when sending words of comfort or encouragement." Her family has "the letters and notes she sent to us over the years," adding "they were what kept her alive." Mrs. Swain died in September 2003.

Taylor, Nancy
Nancy Taylor lives in San Jose, California. "I have two beautiful daughters—Jennifer Lynn, thirty-one, and Diane Marie, twenty," she writes. "The saddest day, November 22, will always live in my heart, the death of hope."

Thornhill, Mrs. J. M.
Freda Mae Thornhill was born in 1912. She was a housewife and loved to paint. She spent several months in a painting school and left behind a collection of artwork that her son and her granddaughter continue to enjoy. She was very politically active and remained involved in politics after the death of President Kennedy. She died in Dallas in 1980.

Tierney, Bridget
Bridget Tierney is an Episcopal priest, serving a congregation in New Lenox, Illinois. She reports that she is "still an ardent Democrat." She worked on Barack Obama's campaign and was in Grant Park on election night—"one of the thrills of my life."

Tippit, Mrs. J. D.
Marie and J. D. Tippit were married for eighteen years and had three children when Officer Tippit was killed on November 22. Mrs. Tippit remarried in 1967.

Tomaro, Dominic A. and family
Mrs. Tomaro reports that she is "now eighty-five-plus years, recently widowed after a sixty-year marriage to Dominic A. Tomaro, another grave loss. I was really on a sixty-year honeymoon! I am a registered nurse—graduated in 1945. Still applying (however not working) for Ohio nurses license to 'keep in touch.'"

Tomashek, Mrs. William
"My mother was a nurse for many years," Mrs. Tomashek's son William writes. "She enjoyed bowling and loved animals. I know that my father and mother both had very high hopes for where the Kennedys could take this nation. I wish that we all had back many of the things that we had in the '60s."

Toomey, Larry
Larry Toomey served with the U.S. Army Signal Corps. He is a civilian employed by the U.S. government. Mr. Toomey enjoys researching the Civil War.

Touchet, John L.
John Touchet was a mechanical engineer. He had four children and earned a pilot's license later in life. He is now eighty-six years old but remembers clearly the night of the Kennedy assassination, when he stood in for a friend tending bar.

Tyler, Diana
Diana Tyler writes: "I am not sixty years old and recently retired after thirty years at USPS."

Van Dyke, Ed
Edward Van Dyke was born in September 1916 in Orange, Essex County, New Jersey. He is a graduate of Choate School, where he was a classmate of John F. Kennedy. He was a businessman and died in September 1986 in Philadelphia, Pennsylvania.

Vrabel, Mr. and Mrs. Andrew
Andrew was born in 1894, and his wife, in 1902. They settled in Bessemer, Pennsylvania, where Mr. Vrabel and his sons all worked in the same plant. The Vrabels had three boys and three girls. Mrs. Vrabel enjoyed sewing and knitting and was, her eighty-two-year-old son recalls, a "staunch Democrat." Mr. Vrabel died in 1980, and his wife died thirteen years later.

Wade, Mr. and Mrs. Aubrey
The Wade family had a small farm and raised tobacco, corn, and cows. There were six children—four boys and two girls. Mrs. Aubrey Wade attended Normal School and emphasized to her children the importance of good penmanship. The Wades' daughter Brenda W. Perdue actually wrote the condolence letter for her mother.

Watson, S/Sgt. and Mrs. William B., Jr.
William B. Watson was killed in Vietnam at An Khe on February 23, 1966, just after his twins' second birthday. Mrs. Watson raised the Watsons' four children on her own, returning to school and working her way through college. She is now retired and lives in New York. The twins live in the South.

Weir, Russell
Mr. Weir felt terrible to the end of his days about what happened to John F. Kennedy. He expressed concern that so few young people seemed to have a keen awareness of the history of this country. He died only recently.

Wentworth, Marcy
Marcy Wentworth writes: "The events in Dallas in 1963 were the defining moment that changed me from a naive schoolgirl into a focused and earnest adult, intent on living my ideals. I completed my university degrees and continue to pursue a lifelong career as teacher and reading specialist. After much travel, I again reside in Austin, Texas, with my geographer husband, near our children and grandchildren."

Whitt, Draper, Lucille, Nelson, and Steve
Draper Whitt was a veteran of WWII who was left disabled outside of France (a bullet entered the base of his skull and exited his ear). Despite his disability, he held

down several jobs prior to his death in 1972. He moved from North Carolina to Virginia after the war to be closer to outpatient treatments for his war injuries. He was a strong advocate for disabled veterans, volunteering at the VA hospital.

Wiggs, Kenneth R., Jr.

Kenneth Wiggs writes: "At the time I wrote this letter I was being trained to work on electronic military fighter aircraft. I stayed for my full four-year term in the military. I also served in Vietnam during this time. After leaving the service, I worked for General Dynamics in Fort Worth, Texas, for about three and a half years before moving into flight simulation for American Airlines. I worked there until I retired in 2002. I still think that Jackie Kennedy is one of the finest women to ever be a First Lady of our nation. She showed a great deal of class and fortitude in her life. She also raised some very fine children to carry the Kennedy name. . . . For a young lady to be thrown into history so violently, she persevered very well and showed how dynamic she could be in her and America's troubled times. . . . The letter was from the heart of a very anguished airman who was seeing a very troubled part [of] history in our country unfold."

Wildesen, Dora A.

Dora Wildesen was born in 1890 in West Virginia. She married and lived on a farm in Maryland for nearly sixty years. She and her husband had one son. Mrs. Wildesen was a housewife and took care of the animals, including hens, sheep, and calves. Her grandson recalls that she was a fantastic bread maker. Every day, at every meal, they would have some sort of hot baked good—bread, rolls, cake, etc. He also remembers helping to take care of the sheep in the summertime in their mountain pasture. His grandmother would make him a sandwich, give him a bottle of water, and he would go off on the pony into the mountain for hours—"it was such a different time then." Their closest neighbor lived a mile away. One of her grandson's early memories is of his grandmother rocking him for hours in a wooden chair. Dora Wildesen died in October 1977.

Wiley, Mrs. John

Mrs. John C. Wiley lived and worked in Washington, D.C., after 1963. She remarried and moved to New England with her husband, a physician. They raised three children and now have four grandchildren. Patricia Parks Wiley Kelleher has worked for a hospice program and as a director of a preschool. "Looking back over these years," she writes, "I am filled with gratitude."

Williams, Mrs. Bessie

Bessie Williams's son writes that his mother "was a widow and mother of fifteen children who worked as a domestic laborer most of her life, was able to purchase a home in

1971 with a whopping $845 down payment. She passed away in 1983." Her son now owns and lives in this property. Bessie, he concludes, "was truly a child of God'!!"

Williams, Ethel C.

Ethel Chavis Williams was born in rural Columbia, South Carolina, in 1920. She completed fifth grade and then continued her education on her own. She moved to Detroit in 1948. She retired from the Detroit Board of Education food services in 1980. She currently lives in senior housing a few blocks from the Messiah Church where she taught and mentored hundreds of children, sang in the choir, welcomed Vietnamese refugees, and hand-addressed thousands of communications to college students and others around the globe. She wrote a poem in 1964 that recollected her 1962 encounter with John F. Kennedy. She told a friend that "returning from day labor she requested to be dropped off near the intersection of Michigan Avenue and West Grand Boulevard, where she surmised the Presidential motorcade would pass. She was highly excited and greatly bemused that she was the only one there. She recalls the cars approaching so very slowly and her amazement at finding herself within five feet of the President. Then I just walked on home, she declares with an open arm shrug." In her poem "Five Feet of Space in an Instant of Time," she expresses her joy at seeing President Kennedy and her regret that she never got to shake his hand. The poem concludes: "But if GOD, whose abode, is in Heaven, Let me cross the Great Divide. I shall find J. F. K. and say to him, Thanks for giving me back my pride. His light has not, gone out in the world, It burns bright, in The Eternal Flame, So that in, generations to come, All shall remember his name. John Fitzgerald Kennedy, Upon whose shoulders, all of our hopes were borne, So long, goodbye, and GOD bless you, A good journey, fare well, and welcome home." Mrs. Williams is eighty-nine years old.

Wise, Blanche

Blanche Wise was born in 1891. She and her husband raised nine children in Enon, Pennsylvania. She died in September 1978, at the age of eighty-six.

Wise, Joyce

Joyce Wise lives in Pennsylvania.

Wofford, Harris

Born in 1926, Harris Wofford graduated from the University of Chicago, Yale, and Howard University Law School. A veteran of World War II, he supported the civil rights movement and was an attorney for the U.S. Civil Rights Commission in the late 1950s. He was a central figure in the Kennedy administration's civil rights policy group, serving as a special assistant to the President. He helped to form the Peace Corps and served as its associate director from 1962 to 1966. He was president of the College at Old Westbury from 1966 to 1970 and of Bryn Mawr College from 1970 to 1978. He served in the U.S. Senate from 1991 to 1995. He currently resides in Washington, D.C.

Wood, Robert L.
Mr. Wood was a longtime officer of the Eastern Star fraternal organization. He belonged to the Methodist church in Dallas. He died in 2000.

Woodrick, Mrs. Riley
Mrs. Woodrick had one daughter. Her nephew recalls that his aunt "was always writing letters to governors and all kinds of people." She had a high school education. She and her husband were farmers. Mrs. Woodrick died in 1982.

Wysota, George
George Wysota lives on Long Island. He has been married for over twenty years and has a daughter who is almost sixteen. He started his own pool business. He notes that he was profoundly moved by the Kennedys and followed the family's activities for years.

Young, Monroe, Jr.
Monroe Young attended college for two years, married, and had two children. His daughter is a schoolteacher and his son is a barber. Monroe died in September 1990 in Dallas at the age of thirty-four.

Zarnowitz, Jill
Jill Zarnowitz graduated from the University of Washington, where she earned a degree in wildlife science. She and her husband now own the Stag Hollow vineyard and winery and live in northwest Oregon. She remembers that she was in the sixth grade when President Kennedy was assassinated and comments that "his and Jackie Kennedy's lives touched me deeply." She became involved in the civil rights movement "in part as a result of President Kennedy's policies."

Zemeski, Robert W.
Robert Walter Zemeski was born in 1945 and grew up in Burlington, New Jersey. After graduating from high school, he joined the Marines in 1963. He received an honorable discharge a year later. He attended Hyles Anderson Bible College in Indiana and earned advanced seminary degrees. He and his wife have four children and nine grandchildren. He has served as a Baptist missionary pastor and has started churches in Ireland since 1979.

Zepp, Fred R.
Fred Zepp was a writer, and his son recalls that his father was not a supporter of President Kennedy's. He remembers calling home the night of November 22 "specifically to speak to my father in search of comfort and reassurance, only to find that I was totally unable to speak to him—all I could do was cry for us all. I shall always

wonder now if my call had any influence on my father's most kind words to Mrs. Kennedy." Mr. Zepp died in 1994.

Zimmerman, Elisabeth
Elisabeth Zimmerman taught English and history before retiring. She has always enjoyed, she writes, "literature and writing" and is considering writing a children's book. She is married and has two sons and six grandsons. She vividly recalls writing as a twelve-year-old to Mrs. Kennedy.

Zinn, Mr. and Mrs. Tempest C.
After writing to Mrs. Kennedy, Nellie Zinn continued living with her husband on a farm in Stewartsville until her failing health forced her into a home. She died at the age of ninety.

Letter Citations

All letters are from The Papers of John F. Kennedy: Condolence Mail, *in the John F. Kennedy Presidential Library, Boston, Massachusetts.*

They have been transcribed as they were found in the archives, with spelling, punctuation, and grammar retained. A few very long letters have had short passages omitted—which do not change the meaning of the originals. Such omissions are indicated in the text by ellipses. Dates are as provided in the letter. When letters are undated, a postmarked date, if available, is signified here by an asterisk [*].

Specific locations are noted below:

114   Bentley, Mary W., November 27, 1963: Adult Letters, box 6, folder 44.

251   Berkery, Margaret W., January 11, 1964: Adult Letters, box 13, folder 103.

67   Bethell, Thomas N., November 25, 1963: Adult Letters, box 17, folder 133.

157   Bingle, Mrs. Charles & family, No Date: Adult Letters, box 6, folder 42.

269   Blumberg, Lisa Booth, No Date: Adult Letters, box 14, folder 104.

244   Boling, Patricia Lee Rita Mary, May 28, 1964: Children's Letters, box 61, folder 36.

261   Bond, Beatrice Joan (Mrs. Wm. Leslie Bond), January 2, 1964: Adult Letters, box 19, folder 151.

114   Boner, Mrs. Linnie, January 15, 1964: Adult Letters, box 17, folder 134.

102   Borders, Mrs. Frank (Vivian), January 17, 1964: Adult Letters, box 8, folder 58.

279   Bradbury, Teresa, December 13, 1963: Adult Letters, box 4, folder 27.5.

209   Bradley, Mrs. Joseph, Jr., November 27, 1963: Adult Letters, box 2, folder 10.

147   Brenner, Peter, December 2, 1963: Adult Letters, box 1, folder 2.

139   Brooks, Mrs. Wilma, November 22, 1964: Adult Letters, box 13, folder 100.

81   Buerman, Fred, November 28, 1963: Adult Letters, box 11, folder 82.

131   Burke, Roger L., Jr., November 22, 1963: Adult Letters, box 6, folder 45.

103   Burril, Mrs. Andrew, November 22, 1964: Adult Letters, box 13, folder 100.

217   Cannon, Irene, April 12, 1964: Adult Letters, box 1, folder 5.

250   Carriker, C. O., November 25, 1963: Adult Letters, box 3, folder 19.

116   Cates, Robert T. M.D., December 2, 1963: Adult Letters, box 8, folder 64.

161   Chapa, Minerva, January 13, 1964: Children's Letters, box 67, folder 81.

259   Chapman, Peg, February 25, 1964: Adult Letters, box 20, folder 155.

165   Charley, Mrs. Pearl A., November 23, 1963: Adult Letters, box 9, folder 66.

182   Clarke, Mr. & Mrs. John W. (Lynne), December 4, 1963: Adult Letters, box 11, folder 81.

234   Cohan, Mrs. Donald S., January 11, 1964: Adult Letters, box 4, folder 27.5.

150   Cohee, Mr. and Mrs. Ralph & family, December 5, 1963: Adult Letters, box 12, folder 90.

163   Cook, Mrs. John G. & Mother Mrs. Henry Wood, January 14, 1964: Adult Letters, box 20, folder 156.

13   Crabtree, Janice (Mrs. W. C.), May 28, 1964: Adult Letters, box 21, folder 164.

18   Cuchia, Mr. & Mrs. Frank, No Date: Adult Letters, box 10, folder 76.

118   Daniel, W. J. B., December 2, 1963: Adult Letters, box 2, folder 12.

266   Davenport, Marilyn, *August 28, 1964: Adult Letters, box 20, folder 154.

99   Davis, Cornelia M., November 27, 1963: Adult Letters, box 5, folder 39.

111   Davis, George T., November 25, 1963: Adult Letters, box 9, folder 68.

273   Davis, Tony & Mrs. Donald K., No Date: Children's Letters, box 58, folder 14.

186   Delaney, Henry H., December 11, 1963: Adult Letters, box 22, folder 175.

179   Diamond, Ellen, No Date: Adult Letters, box 13, folder 102.

260   Diaz, Alden, November 22, 1963: Adult Letters, box 5, folder 35.

23   DiGeorgio, Susan, November 22, 1963: Adult Letters, box 11, folder 84.

175   Donnally, Mrs. Emma, *December 14, 1963: Adult Letters, box 7, folder 49.

277   Dryden, Jane, January 18, 1964: Children's Letters, box 65, folder 67.

255   Dumais, Roland, February 14, 1964: Personal Remembrances, box 1, folder 9.

227   Emery, Harry T., No Date: Adult Letters, box 12, folder 94.

190   Emmitt, Tom, December 3, 1963: Adult Letters, box 1, folder 6.

236   Ernst, Pauline, Katherine, William & Douglas, November 30, 1963: Adult Letters, box 14, folder 104.

98   Evers, Mrs. Medgar, November 23, 1963: VIP Letters, box 2, folder E.

239   Farquhar, Donald, No Date: Adult Letters, box 2, folder 12.

42   Fiola, Mr. & Mrs. Roland A., Lisa, Monique, Michelle & Je'Neanne. November 23, 1963: Adult Letters, box 7, folder 52.

124   Floodas, Mr. & Mrs. John J., *January, 16 1964: Unprocessed Letters, box 2.

189   Fowler, Dona, *February 3, 1964: Adult Letters, box 19, folder 146.

168   Gambardelli, Arthur, November 29, 1963: Adult Letters, box 9, folder 65.

47   Garfield, Rudolph H., November, 1963: VIP Letters, box 2, folder GA

221   Gatewood, Chris K., January 28, 1964: Adult Letters, box 2, folder 14.

83   Geist, Mrs. Carroll A., January 16, 1964: Unprocessed Letters, box 2.

146   Gidion, Gabriele, November 28, 1963: Adult Letters, box 4, folder 27.5.

154   Givens, Ruby, December 3, 1963: Adult Letters, box 14, folder 110.

216   Glassner, Mary, *January 25, 1964: Adult Letters, box 3, folder 20.

11   Glimpse, Nancy, Kenneth, Rick, Brandon & Perry, November 22, 1963: Adult Letters, box 21, folder 165.

36   Golub, Mrs. Shirley, November 25, 1963: Adult Letters, box 4, folder 30.

62   Gonzales, Henry, December 8, 1963: Adult Letters, box 1, folder 6.

272   Grumblatt, Douglas K., *November 25, 1963: Children's Letters, box 61, folder 38.

46   Gum, Perry C., December 1, 1963: Adult Letters, box 21, folder 164.

231   Hammonds, Mr. & Mrs. R.T., December 11, 1963: Unprocessed Letters, box 3.

86   Hanrahan, Stephen J., December 2, 1963: Adult Letters, box 9, folder 70.

134   Harayda, Eileen, March 16, 1964: Adult Letters, box 13, folder 100.

19   Harper, Mr. & Mrs. J., December 17, 1963: Adult Letters, box 15, folder 120.

224   Harris, Charles A., November 27, 1963: Personal Remembrances, box 3, folder 24.

262   Harris, Irene, March 6, 1964: Adult Letters, box 8, folder 59.

209   Hebb, Jo, December 2, 1963: Adult Letters, box 20, folder 162.

28   Hemmerle, Patricia Anne, January 20, 1964: Children's Letters, box 64, folder 55.

75   Henrikson, Mike, No Date: Adult Letters, box 14, folder 108 (attached to letter from Melvin C. Parrish).

195   Hesch, Mildred F. (Mrs. Robert L.), November 23, 1963: Adult Letters, box 16, folder 122.

212   Hilliard, Jerrine, December 25, 1963: Unprocessed Letters, box 3.

264   Hirsch, Janis, January 16, 1964: Children's Letters, box 59, folder 23.

242   Hoerl, Norbert A., December 4, 1963: Adult Letters, box 2, folder 11.

245   Holman, Clinton Hale, December 6, 1963: Adult Letters, box 3, folder 20.

170   Housley, Mrs. Lennie Gore, *March 1964: Adult Letters, box 16, folder 127.

101   Hughes, Langston, November 22, 1963: VIP Letters, Box 2, HU.

106   Jackson, Katherine Dowd, November 23, 1963: Adult Letters, box 2, folder 11.

44   Jackson, Larry, November 22, 1963: Adult Letters, box 12, folder 90 (attached to letter from Mrs. Whitley Ray).

49   Jacobs, Mrs. Howard A. & family, November 27, 1963: Adult Letters, box 9, folder 66.

265   Jakusik, John H., No Date: Adult Letters, box 12, folder 90.

275   Jones, Sandi, September 4, 1964: Children's Letters, box 60, folder 28.

278   Junker, Ellen, *November 27, 1963: Children's Letters, box 60, folder 25.

64   Katzberg, Mrs. G., January 14, 1964: Adult Letters, box 7, folder 48.

130   Keat, Betty (Mrs. James), November 22, 1963: Adult Letters, box 11, folder 84.

71   Keegin, Arch C. April 22, 1964: Adult Letters, box 3, folder 24.

42   Keller, The Rev. David, November 22, 1963: Adult Letters, box 21, folder 163.

191   Kendler, Karen S., January 14, 1964: Adult Letters, box 22, folder 175.

248   Kilmurry, Ellen, December 21, 1963: Children's Letters, box 62, folder 42.

276   Klemkosky, Brenda, January 24, 1964: Children's Letters, box 57, folder 2.

129   Koop, Mrs. Patricia, November 22, 1963: Adult Letters, box 8, folder 56.

275   Lane, Suzan Elizabeth, December 6, 1963: Children's Letters, box 65, folder 64.

69   Legg, Jim, Jr., November 25, 1963: Adult Letters, box 10, folder 73.

262   Lewis, Lillie M., November 23, 1963: Adult Letters, box 6, folder 46.

241   Link, Leonard L., Jr., No Date: Personal Remembrances, box 1, folder 2.

183   Lockeby, Janis M., *January 16, 1964: Adult Letters, box 1, folder 6.

79   Lodge, George C., November 26, 1963: VIP Letters, box 3, folder LO.

288   Longeneker, Grace, December 6, 1963: Adult Letters, box 7, folder 55.

200   Louk, R.R., February 14, 1964: Adult Letters, box 20, folder 161.

54   Lounsbery, Anna (Mrs. John), Christmas, 1963: Adult Letters, box 16, folder 128.

54   Lounsbery, Annie, November 23, 1963: Adult Letters, box 16, folder 128 (attached to letter from Anna Lounsbery).

29   Lowrey, Irene, December 7, 1963: Adult Letters, box 11, folder 83.

185   Lundstrom, Gretchen, November 26, 1963: Adult Letters, box 11, folder 87.

287   Lynch, Sheila J., January 19, 1964: Adult Letters, box 8, folder 58.

223   MacArthur, General, November 22, 1963: VIP Letters, box 3, folder MAA-MAC.

233   Macho, Mary and Adolph Sr., January 21, 1964: Adult Letters, box 1, folder 8.

26   Mackey, Vivian, November 24, 1963: Adult Letters, box 5, folder 39.

167   Manfre, Mrs. Pati and Vivian, No Date: Adult Letters, box 17, folder 129.

81   Martin, E.G., November 28, 1963: Adult Letters, box 8, folder 57.

225   Matteson, Austin R. & family, November 25, 1963: Adult Letters, box 16, folder 122.

22   McClain, David Blair, November 22, 1963: Personal Remembrances, box 1, folder 11.

284   McCormack, M.R. (Mrs. Melbro R.), November 26, 1963: Adult Letters, box 6, folder 42.

102   McKenney, Daisy H., No Date: Adult Letters, box 15, folder 117.

20   McLean, Margaret, January 17, 1964: Adult Letters, box 7, folder 53.

246   McManus, Mrs. Dorothy, January 26, 1964: Adult Letters, box 22, folder 176.

24   McMillen, Mary, No Date: Children's Letters, box 66, folder 70.

x   McMurtry, Mrs. Gertrude, November 28, 1963: Adult Letters, box 11, folder 88.

230   McNeil, Grady, *July 30, 1964: Adult Letters, box 18, folder 143.

258   Meader, Vaughn, No Date: VIP Letters, box 3, folder ME.

55   Melder, Mrs. M. F., January 16, 1964: Adult Letters, box 22, folder 176.

56   Melder, Ray (Lt. Raymond A.), November 26, 1963: Adult Letters, box 22, folder 176 (attached to letter from Mrs. M.F. Melder).

208   Mesaros, Mrs. Margaret, January 2, 1964: Adult Letters, box 7, folder 50.

49   Metzger, Mrs. Regina, November 23, 1963: Adult Letters, box 8, folder 59.

135   Michalski, Lorraine J., December 4, 1963: Adult Letters, box 16, folder 127.

135   Michalski, Tom, No Date: Adult Letters, box 16, folder 127 (attached to letter from Lorraine J. Michalski).

213   Milano, Helen M., January 13, 1964: Adult Letters, box 7, folder 49.

59   Mitchell, Eileen, November 22, 1963: Adult Letters, box 12, folder 90.

109   "Mrs. American Citizen," November 28, 1963: Unprocessed Letters, box 1.

41   Nash, Mrs. Frances, November 23, 1963: Adult Letters, box 20, folder 157.

107   "A Negro Who beleave In God," No Date: Adult Letters, box 7, folder 54.

267   Nichols, Linda Gayle & Donnie, December 18, 1963: Adult Letters, box 7, folder 48.

160   Nichols, Ruby K., November 25, 1963: Adult Letters, box 17, folder 133.

109   Nies, Mary F. (Mrs. Frederick J.), January 17, 1964: Adult Letters, box 18, folder 138.

18   Oakey, Carol, January 6, 1964: Adult Letters, box 11, folder 82.

193   O'Connor, Captain Peter N., November 25, 1963: Adult Letters, box 1, folder 7.

103   Oglesby, Susie, December 17, 1963: Adult Letters, box 16, folder 122.

162   Oldham, Mrs. Helen, November 25, 1963: Adult Letters, box 9, folder 65.

40   Ostby, Janeen, February 4, 1964: Adult Letters, box 10, folder 73.

26   Palumbo, Jo-Ann, November 23, 1963: Children's Letters, box 63, folder 49.

75   Parrish, Melvin C., January 23, 1964: Adult Letters, box 14, folder 108.

115   Parsons, Catherine B. (Mrs. J.S.), No Date: Adult Letters, box 3, folder 22.

264   Paul, Mrs. Jerome, No Date: Adult Letters, box 7, folder 49.

222   Pinkney, Grace, December 5, 1963: Adult Letters, box 5, folder 39.

256   Poberezny, Mrs. Thomas (Doris), January 18, 1964: Unprocessed Letters, Box 2.

271   Pond, Kate, No Date: Children's Letters, box 57, folder 2.

205   Pucka, Elizabeth and Becky, November 23, 1963: Adult Letters, box 6, folder 43.

143   Quan, Mary, April 4, 1964: Adult Letters, box 8, folder 59.

274   Radell, Kevin, *December 19, 1963: Children's Letters, box 66, folder 74.

166   Ralston, Frances, December 6, 1963: Adult Letters, box 4, folder 28.

220   Rassche, Mrs. Catherine, July 17, 1964: Adult Letters, box 8, folder 61.

43   Ray, Mrs. Whitley, November 29, 1963: Adult Letters, box 12, folder 90.

38   Rice, Leonard C., November 23, 1963: Adult Letters, box 8, folder 60.

270   Rimer, Barbara, January 17, 1964: Unprocessed Letters, box 2.

267   Robinson, Ethel M., December 7, 1963: Adult Letters, box 12, folder 94.

80   Robinson, Mr. and Mrs. Hugh B., Jr., December 1, 1963: Adult Letters, box 12, folder 94.

86   Rock, Leonard F., November 28, 1963: Adult Letters, box 3, folder 24.

41   Rosenberg, Martin, No Date: Adult Letters, box 6, folder 44.

108   Ross, Martha, November 28, 1963: Adult Letters, box 4, folder 29.

127   Runnals, Thomas H., November 25, 1963: Adult Letters, box 12, folder 93.

12   Russell, J.E.Y., December 1, 1963: Adult Letters, box 21, folder 165.

229   Sanders, Donald W., January 21, 1964: Adult Letters, box 18, folder 138.

239	Schechter, Mrs. Jean, April 29, 1964: Adult Letters, box 2, folder 15.
194	Schwen, Marcia, November 28, 1963: Adult Letters, box 17, folder 134.
101	Scott, Mrs. E., November 23, 1963, Adult Letters, box 22, folder 179.
256	Seiler, Ira, November 28, 1963: VIP Letters, box 4, folder SE.
142	Sheldon, Judy, December 4, 1963: Adult Letters, box 4, folder 29.
285	Sherman, Estelle, December 5, 1963: Adult Letters, box 12, folder 89.
169	Siegel, Natalie, November 24, 1963: Adult Letters, box 4, folder 30.
253	Silverstein, Irving, November 28, 1963: Personal Remembrances, box 2, folder 16.
120	Simrin, Arlene & family, November 27, 1963: Personal Remembrances, box 1, folder 1.
207	Skeats, Claudine R. (Mrs. A.E.), November 29, 1963: Personal Remembrances, box 3, folder 28.
153	Smith, Ernan H., December 25, 1963: Adult Letters, box 1, folder 2.
197	Smith, Mrs. Paul F., January 16, 1964: Adult Letters, box 9, folder 66.
17	Smith, Tommy, No Date: Children's Letters, box 59, folder 18.
165	Snell, Mrs. Alma, November 23, 1963: Adult Letters, box 22, folder 167.
173	Snider, Mrs. Merlene, No Date: Adult Letters, box 12, folder 93.
123	Sooby, Donna, December 2, 1963: Adult Letters, box 9, folder 67.
32	South, Mary, November 24, 1963: Adult Letters, box 5, folder 33.
83	Spector, Pauline & Sol K., January 20, 1964: Adult Letters, box 1, folder 8.
215	Stafford, Mrs. Ruth, No Date: Adult Letters, box 7, folder 50.
52	Stamos, Katherine, January 25, 1964: Adult Letters, box 8, folder 64.
52	Stamos, Spiro, November 22, 1963: Adult Letters, box 8, folder 64 (attached to letter from Katherine Stamos).
206	Stanley-Brown, Katharine, November 24, 1963: Adult Letters, box 17, folder 129.
232	Starr, Morris, November 29, 1963: Adult Letters, box 2, folder 16.
181	Steinhart, John, November 23, 1963: Adult Letters, box 2, folder 11.
90	Stone, Helen, December 3, 1963: Adult Letters, box 14, folder 107.
127	Storll, Mrs. F., January 17, 1964: Adult Letters, box 14, folder 104.
105	Swain, Mrs. Marzell, December 2, 1963: Adult Letters, box 15, folder 114.
31	Taylor, Nancy, *December 8, 1963: Children's Letters, box 57, folder 1.
63	Thornhill, Mrs. J. M., February 7, 1964: Adult Letters, box 11, folder 82.
281	Tierney, Bridget, January 16, 1964: Personal Remembrances, box 1, folder 9.
206	Tippit, Mrs. J. D., November 24, 1963: VIP Letters, box 4, folder Tippit, Mrs. J.
233	Tomaro, Dominic A. & family, January 16, 1964: Adult Letters, box 7, folder 49.
73	Tomashek, Mrs. William, November 25, 1963: Adult Letters, box 17, folder 135.

11	Toomey, Larry, November 22, 1963: Adult Letters, box 17, folder 132.
126	Touchet, John L., November 24, 1963: Adult Letters, box 20, folder 155.
270	Tyler, Diana, No Date: Children's Letters, box 59, folder 18.
254	Van Dyke, Ed, January 14, 1964: VIP Letters, box 4, folder V.
263	Vrabel, Mr. & Mrs. Andrew, August, 1964: Personal Remembrances, box 1, folder 2.
73	Wade, Mr. and Mrs. Aubrey & family, January 13, 1964: Adult Letters, box 9, folder 65.
xvi	Watson, Staff Sargeant, & Mrs. William B., Jr., December 16, 1964, Adult Letters, box 18, folder 144.
130	Weir, Russell E., November 27, 1963: Adult Letters, box 12, folder 92.
61	Wentworth, Marcy, November 25, 1963: Adult Letters, box 11, folder 87.
226	Whitt, Draper, Lucille, Nelson & Steve, November 22, 1963: Adult Letters, box 15, folder 119.
60	Wiggs, Kenneth R., Jr., November 22, 1963: Adult Letters, box 1, folder 8.
176	Wildesen, Dora A., March 17, 1964: Adult Letters, box 8, folder 61.
210	Wiley, Mrs. John J., December 20, 1963: Adult Letters, box 6, folder 121.
155	Williams, Mrs. Bessie, February 10, 1964: Adult Letters, box 5, folder 38.
100	Williams, Ethel C., November 29, 1963: Adult Letters, box 17, folder 133.
211	Wise, Blanche, December, 1963: Unprocessed Letters, box 2.
153	Wise, Joyce, January 16, 1964: Unprocessed Letters, box 2.
287	Wofford, Harris, November 26, 1963: VIP Letters, box 4, folder WO-WY.
58	Wood, Robert L., November 22, 1963: Adult Letters, box 10, folder 79.
151	Woodrick, Mrs. Riley (Earlene), March 4, 1964: Adult Letters, box 1, folder 5.
280	Wysota, George, November 27, 1963: Children's Letters, box 59, folder 17.
243	Young, Monroe, Jr., December 1, 1963: Children's Letters, box 63, folder 47.
273	Zarnowitz, Jill, November 28, 1963: Children's Letters, box 57, folder 4.
188	Zemeski, Pvt. Robert W., November 23, 1963: Adult Letters, box 5, folder 36.
48	Zepp, Fred R., November 22, 1963: Adult Letters, box 5, folder 36.
33	Zimmerman, Elisabeth, January 5, 1964: Children's Letters, box 65, folder 64.
138	Zinn, Mr. and Mrs. Tempest C., No Date: Adult Letters, box 21, folder 170.

Notes

Original spelling and grammar is retained in all quoted archival documents. Editor's clarifications and additions are indicated in brackets.

Letters are from The Papers of John F. Kennedy: Condolence Mail, *in the John F. Kennedy Presidential Library, Boston, Massachusetts, cited here as "Condolence Mail, JFKL."*

The interview of Nancy Tuckerman and Pamela Turnure recorded by Mrs. Wayne Fredericks, 1964, John F. Kennedy Library Oral History Program, cited here as "Tuckerman, Turnure OH, JFKL."

Introduction

xi "Katherine Dowd Jackson sat down": author's interview with Edna Jackson-Goins, October 2009. "Dear beloved one": Katherine Dowd Jackson to Mrs. John F. Kennedy, November 23, 1963, Adult Letters, box 11, folder 84, Condolence Mail, JFKL.

xi "What can anyone say?": Mrs. Regina Metzger to Mrs. John F. Kennedy, 23 November 1963, Adult Letters, box 8, folder 59, Condolence Mail, JFKL. "As no other First Family": Cornelia Burge to Mrs. John F. Kennedy, Thanksgiving Day 1963, Children's Letters, box 65, folder 61, Condolence Mail, JFKL.

xii "Surely this generation": Mrs. Paul F. Smith to Mrs. John F. Kennedy, January 16, 1964, Adult Letters, box 9, folder 66, John F. Kennedy Condolence Mail, JFKL.

xii Impact of the Kennedy assassination: A 1999 study by the Pew Research Center found that the Kennedy assassination was "not only the earliest event that a majority of Americans can still recall, it is also the most potent American memory for those who lived through it. Nine out of ten Americans who were old enough at the time say they remember exactly what they were doing when they heard the news of the 35th president's death." Among Americans then aged 55 to 64, "98% remember exactly what they were doing when they heard the news of his assassination." Pew Research Center for the People and the Press, "America's Collective Memory," in *Technology Triumphs, Morality Falters*, July 3, 1999, p. 283, http://people-press.org/report/57/technology-triumphs-morality-falters.

xiii   "How does a nobody write": Mrs. Wilbert George to Mrs. John F. Kennedy, November 26, 1963, Adult Letters, box 19, folder 150, Condolence Mail, JKL. "The following Monday, mail delivery": Author's interview with Nancy Tuckerman, September 1, 2009; Tuckerman, Turnure OH, JFKL, pp. 42–48. "On one occasion": Author's interview with Nancy Tuckerman, September 1, 2009.

xiv   "the first time in 15 years of television": "Millions Watch Oswald Killing on 2 Networks," *New York Times*, November 25, 1963, pp. 1, 10.

xiv   "a million people stood": "Million in Capital See Cortege Roll on to Church and Grave," *New York Times*, November 26, 1963, p. 1.

xv   "letters to Hyannisport": "Kennedy's House Lashed By Storm," *New York Times*, December 1, 1963, p. 57. On the scale of condolence letters, see: "Mrs. Kennedy Thanks 800,000 Who Expressed Their Sympathies," *New York Times*, January 15, 1964, pp. 1, 13. Finding Aid, "The Papers of John F. Kennedy: Condolence Mail," JFKL.

xvii   "House Appropriations Committee": "House Panel Asks $50,000 For Staff of Mrs. Kennedy," *New York Times*, May 7, 1965, p. 28; Tuckerman, Turnure OH, JFKL, pp. 44–45.

xvii   "a letter to the President's daughter": Mrs. Jeri C. Schleunes to Mrs. John F. Kennedy, November 22, 1963, Adult Letters, box 13, folder 102, Condolence Mail, JFKL. "Texan John Titmas": John Titmas Photograph of Mr. and Mrs. John F. Kennedy, PX 91-32, Audio-Visual Collections, JFKL.

xviii Jacqueline Kennedy's television appearance: "Mrs. Kennedy Thanks 800,000 Who Expressed Their Sympathies," *New York Times*, January 15, 1963, pp. 1, 13. Jacqueline Bouvier Kennedy, "Address Thanking Everyone for Kind Wishes," Videotaped Recording, 1 FP: 6, JFKL.

xx   On handling of condolence mail: Author's interview with Nancy Tuckerman, September 1, 2009; Adult Letters, box 22, "Volunteers, Instructions to and Sample Reply Material," Condolence Mail, JFKL; Tuckerman, Turnure OH, JFKL, pp. 45–47. "Margaret Mead sent word": Author's interview with Nancy Tuckerman, September 1, 2009. "As late as 1966": "Items on Kennedy Pour Into Office," *New York Times*, May 1, 1966, p. 119.

xxi   Details on the Condolence Mail collection: Finding Aid, "The Papers of John F Kennedy: Condolence Mail," JFKL. The three boxes of unprocessed letters do not differ significantly from the larger collection of adult and children's letters. Most messages written in November and December of 1963 reside in the general collection.

xxi   "among the most consistently popular of Presidents": "Presidential Approval Ratings," http://www.ropercenter.uconn.edu/data_access/data/presidential_approval.html; "Presidential Approval Ratings-Gallup Historical Statistics and Trends," http://www.gallup.com/poll/116677/presidential-approval-ratings-gallup-historical-statistics-trends.aspx; Fred I. Greenstein, *The Presidential Difference: Leadership Style from FDR to Barack Obama* (Princeton: Princeton University Press, 2009), pp. 68–69.

xxi   Relationship between the President and the press: Robert Dallek, *An Unfinished Life: John F. Kennedy, 1917–1963* (Boston: Little, Brown, 2003), pp. 375–76, 477–79; Montague Kern, Ralph B. Levering, and Patricia W. Levering, *The Kennedy Crises: The Press, The Presidency, and Foreign Policy* (Chapel Hill: University of North Carolina Press, 1984); Michael R. Beschloss, *The Crisis Years: Kennedy and Khrushchev, 1960–1963* (New York: HarperCollins, 1991), pp. 611–15 and *passim*; Jeffrey E. Cohen, *The Presidency in the Era of 24-Hour News* (Princeton: Princeton University Press, 2008); Andrew Rudalevige, *The New Imperial Presidency: Renewing Presidential Power after Watergate* (Ann Arbor: University of Michigan Press, 2006), pp. 41–42.

xxii   Impact of Vietnam and Watergate: David Kaiser, *American Tragedy: Kennedy, Johnson and the Origins of the Vietnam War* (Cambridge: Harvard University Press, 2000); Richard Reeves, *President Kennedy: Profile of Power* (New York: Simon and Schuster 1993); James T. Patterson, *Grand Expectations: The United States, 1945–1974* (New York: Oxford University Press, 1996).

xxii   Kennedy press conferences: Dallek, *An Unfinished Life*, pp. 335–36. "My husband and I use to get such a kick": Mrs. Riley Woodrick to Mrs. John F. Kennedy, March 4, 1964, Adult Letters, box 1, folder 5, Condolence Mail, JFKL.

xxii   "he also used the press conferences": Dallek, *An Unfinished Life*, pp. 335–36.

xxiii   "Mr. Kennedy taught my children": Josie L. Collins to Mrs. John F. Kennedy, Adult Letters, box 9, folder 68, Condolence Mail, JFKL. "three out of four adults": Dallek, *An Unfinished Life*, p. 336. Public attention to Jacqueline Kennedy: Dallek, *An Unfinished Life*, pp. 478–79. "A Tour of the White House with Mrs. John F. Kennedy," The Museum of Broadcast Communications, http://www.museum.tv/archives/etv/T/htmlT/tourofthew/tour ofthew.htm.

xxiii   "youngest children to the White House": Doug Wead, *All the Presidents' Children* (New York: Atria Books, 2003), p. 22.

xxiii   "a warm and engaging man": Mrs. Edward D. Heath to Mrs. John F. Kennedy, January 25, 1964.

xxiv   "As Jacqueline Kennedy herself noted": Theodore White, Original Hand-Written Notes of "Camelot" Interview with Mrs. Kennedy, Theodore H. White Papers, box 40, Camelot Documents, JFKL. "His death is disquieting to me": Dick Santoro to Barbara Longsworth, enclosed in Barbara Longsworth to Mrs. John F. Kennedy, Personal Remembrances, box 2, folder 13, Condolence Mail, JFKL.

xxv   "We loved your husband": Vivian Borders to Mrs. John F. Kennedy, Adult Letters, box 8, folder 58. "I am colored and poor": Grace Pinkney to Mrs. John F. Kennedy, December 5, 1963, Adult Letters, box 5, folder 39, both in Condolence Mail, JFKL.

xxv   "I am a Florida dairy farmer": Russell Weir to Mrs. John F. Kennedy, November 27, 1963, Adult Letters, box 12, folder 92, Condolence Mail, JFKL. "in two seconds history's course": Vesta I. Nelson to Mrs. John F. Kennedy, November 30, 1963, Adult Letters, box 1, folder 6, all in Condolence Mail, JFKL.

xxv   "This land was ours": Robert Frost, "The Gift Outright," *The Poetry of Robert Frost: The Collected Poems Complete and Unabridged* (New York: Holt, 1979), p. 348. On Kennedy's appearance at Amherst: Amherst College, "The President and the Poet: John F. Kennedy at Amherst, 1963," https://www.amherst.edu/library/archives/exhibitions/kennedy; Arthur M. Schlesinger Jr., *A Thousand Days: John F. Kennedy in the White House* (New York: Greenwich House, 1965), pp. 1015–16. Frost at the Kennedy inauguration: Schlesinger, *A Thousand Days*, pp. 1–3.

xxviii   "The coffin was very small": Jane Townes to Mrs. John F. Kennedy, November 25, 1963, Children's Letters, box 67, folder 76, Condolence Mail, JFKL.

November 22, 1963

3   "History Jumping Up": N. Douglas Paddy to Mrs. John F. Kennedy, December 26, 1963, Adult Letters, box 7, folder 55, Condolence Mail, JFKL. Trip to Texas: William Manchester, *The Death of a President* (New York: Penguin, 1977), pp. 70–87; Robert Dallek, *An Unfinished Life: John F. Kennedy 1917–1963* (Boston: Little Brown, 2003), pp. 691–93. "A woman who waited four hours": Mrs. Eugene E. McCockey to Mrs. John F. Kennedy, 2 January 1964, Unprocessed Letters, box 2, Condolence Mail, JFKL.

4   Events in Fort Worth: Manchester, *Death of a President*, pp. 106–8, 112–14, 116–25; "Kennedy at Fort Worth Kept Darting Into Crowds," *Dallas Morning News*, November 23, 1963, p. 11; "The Final Hours of Kennedy's Life," *New York Times*, November 23, 1963, p. 7; Jeb Byrne, "The Hours Before Dallas: A Recollection by President Kennedy's Advance Man," *Prologue Magazine*, vol. 32, no. 2 (Summer 2000), http://www.archives.gov/publications/prologue/2000/summer/jfk-last-day-3.html; John F. Kennedy, "Remarks at the Breakfast at the Forth Worth Chamber of Commerce," November 22, 1963, http://www.jfklibrary.org/Historical+Resources/Archives/Reference+Desk/Speeches.

4   "roughed up Stevenson": "Stevenson Booed and Hit by Dallas Demonstrators," *New York Times*, October 25, 1963, pp. 1, 6.

5   "black bordered ad": "Welcome Mr. Kennedy to Dallas," American Fact-

Finding Committee, ad run in *Dallas Morning News*, November 22, 1963, copy provided to author by Jim Lehrer; Warren Commission, *Report of the President's Commission on the Assassination of President Kennedy* (Washington, D.C.: Government Printing Office, 1964), vol. 1, chapter 2, p. 40. "We're heading into nut country": Dallek, *An Unfinished Life*, p. 693. "last night would have been": Manchester, *Death of a President*, p. 121.

5   "I told the children": Mrs. Mary Scott to Mrs. John F. Kennedy, November 29, 1963, Adult Letters, box 21, folder 168, Condolence Mail, JFKL.

6   Scene at Love Field: Manchester, *Death of a President*, pp. 128–30. "She had been given yellow roses": Theodore H. White, "For President Kennedy: An Epilogue." *Life*, December 6, 1963, in Theodore H. White Papers, box 59, JFKL. "even though Dallas was mainly": Mary Kay McCallum to Mrs. John F. Kennedy, November 23, 1963, Children's Letters, box 60, folder 31, Condolence Mail, JFKL.

6   "newspapers outlined the motorcade route": Manchester, *Death of a President*, p. 94; Warren Commission, *Report*, chapter 2, p. 40. "Just as your car turned": Ralph G. Falkner to Mrs. John F. Kennedy, December 5, 1963, Adult Letters, box 22, folder 162, Condolence Mail, JFKL.

7   Details on the motorcade: Warren Commission, *Report*, chapter 2, pp. 33–40, 44–48; Manchester, *Death of a President*, pp. 135–37, p. 152. "people stood on the awnings": Janice Crabtree to "Dear Father," November 27, 1963, attached to Janice Crabtree to Mrs. John F. Kennedy, May 28, 1963 [*sic*], Adult Letters, box 21, folder 164, Condolence Mail, JFKL.

8   "One Catholic nun": Simon and Catherine O'Donohue to Mrs. John F. Kennedy, December 3, 1963; Sister Maura to Simon and Catherine O'Donohue, November 28, 1963, letters attached, Personal Remembrances, box 3, folder 23, Condolence Mail, JFKL. "then they were gone": Crabtree to "Dear Father," p. 4.

8   "Jacqueline Kennedy waved": Warren Commission, *Report*, chapter 2, p. 49. "She anticipated the relief": "Testimony of Mrs. John F. Kennedy Before the Warren Commission," in Warren Commission, *Report*, vol. 5, pp. 178–80; Manchester, *Death of a President*, pp. 153–58. "full of blood and red roses": Theodore White, "Handwritten Notes of Camelot Interview with Jacqueline Kennedy," Theodore H. White Papers, box 40, Camelot Documents, JFKL.

8   "Spectators reacted": Manchester, *Death of a President*, pp. 153–58.

9   "Among them was Bob Jackson": Warren Commission, *Report*, chapter 3, p. 65.

9   "felt like an 'eternity'": "Testimony of Mrs. John F. Kennedy Before the Warren Commission," in Warren Commission, Report, vol. 5, p. 180. "news of the assassination attempt began to break": Manchester, *Death of a President*, pp. 167–68, 189–90, 243–44; Erik Barnouw, *A History of Broadcasting in America, Vol. 3: The Image Empire. From 1953* (New York: Oxford University Press, 1970), p. 228; Bob Huffaker, *When the News Went Live: Dallas, 1963* (New York: Taylor Trade Publishing, 2007).

17   "Calls for stretchers": Mrs. Dorothy Smith to Mrs. John F. Kennedy, Smith, December 4, 1963, Personal Remembrances, box 1, folder 1, Condolence Mail, JFKL.

18   "I saw your beloved": Marie Davis to Mrs. John F. Kennedy, January 17, 1964, Unprocessed Letters, box 2, Condolence Mail, JFKL.

20   "I was watching the parade": There was no live television coverage of the motorcade in downtown Dallas. Local radio stations did provide live coverage from reporters' vantage points along the route. The Kennedys' arrival at Love Field was covered live by local television, making use of pool cameras that were also stationed at the Trade Mart where the motorcade was headed. WFAA-TV (Dallas Channel 8) interrupted its regular programming at 12:45 with the first UPI announcement. Its program director, Jay Watson, interviewed eyewitnesses, one young couple still visibly shaking as they recounted what they had just seen. Watson also interviewed Abraham Zapruder, who was filming the motorcade at the moment of the assassination. Huffaker, *When the News Went Live*, pp. 3–9; http://embedr.com/playlist/archives-of-wfaa-tv-dallas-jfk-assassination-footage.

36   "Patients in hospitals": Mrs. Thomas Hurta to Mrs. John F. Kennedy, January 16, 1964, Adult Letters, box 9, folder 65, Condolence Mail, JFKL. "One man dying of cancer": Mrs. Marcella Pieper, November 25, 1963, Adult Letters, box 9, folder 72, Condolence Mail, JFKL.

46   "I was born the first day of April": E. Mae Greene to Mrs. John F. Kennedy, November 27, 1963, Adult Letters, box 20, folder 159, Condolence Mail, JFKL.

46   "I am old enough": Edmund F. Jewell to Mrs. John F. Kennedy, November 30, 1963, Adult Letters, box 22, folder 175, Condolence Mail, JFKL.

57   "the President's body was brought back": Manchester, *Death of a President*, pp. 347–48, 323–24.

58   "She asked herself why": Theodore White, "Handwritten Notes of Camelot Interview with Jacqueline Kennedy," Theodore H. White Papers, box 40, Camelot Documents, JFKL.

58   "Most persons outside the state": Jean Maxwell to Mrs. John F. Kennedy, November 22, 1963, Adult Letters, box 3, folder 23, Condolence Mail, JFKL. "I feel in some way": H. Howard Howard, to Mrs. John F. Kennedy, November 22, 1963, Adult Letters, box 6, folder 41, Condolence Mail, JFKL.

65   "newspapers published in bold headlines": *New York Times*, November 23, 1963, p. 1; *Dallas Morning News*, November 23, 1963, p. 1. "Mrs. Kennedy had returned": Manchester, *Death of a President*, pp. 418–23, 435–39.

66   On the events of Sunday morning: Manchester, *Death of a President*, pp. 656–58; Vincent Bugliosi, *Four Days in November: The Assassination of President John F. Kennedy* (New York: Norton, 2007), pp. 79–169; Gerald Posner, *Cased Closed: Lee Harvey Oswald and the Assassination of JFK* (New York: Random House, 1993), pp. 273–80.

66   "thousands stood in line": "Thousands Pass Bier at Night Despite the Cold and Long Wait," *New York Times*, November 25, 1963, p. 2.

67   "I have been taught all my life": Marsha Hardin to Mrs. John F. Kennedy, January 18, 1964, Unprocessed Letters, box 3, Condolence Mail, JFKL.

Politics, Society, and President, 1963

91   "He was Born Holding a Flag": Bertha Schultz to Mrs. John F. Kennedy, November 22, 1963, Adult Letters, box 7, folder 52, Condolence Mail, JFKL. "It will not be easy": Theodore Sorensen, *Kennedy* (New York: Harper and Row, 1965), p. 757.

92   "It still riles me to think": Bette Douches to Mrs. John F. Kennedy, no date, Adult Letters, box 9, folder 69, Condolence Mail, JFKL. "We may not realize that fault now": Eileen Harayda to Mrs. John F. Kennedy, March 16, 1964, Adult Letters, box 13, folder 100, Condolence Mail, JFKL.

92   "tensions that challenges to segregations had heightened": Taylor Branch, *Parting the Waters: America in the King Years 1954–63* (New York: Simon and Schuster, 1988), p. 880 and chapter 22, *passim*.

93   "A rare letter from a white supremacist": Continent Preservation Party to Mrs. John F. Kennedy, December 7, 1963, Adult Letters, box 19, folder 148, Condolence Mail, JFKL.

93   "The day he was born": Bertha Schultz to Mrs. John F. Kennedy, November 22, 1963, Adult Letters, box 7, folder 52, Condolence Mail, JFKL.

94   On Kennedy and civil rights, see: G. Calvin Mackenzie and Robert Weisbrot, *The Liberal Hour: Washington and the Politics of Change in the 1960s* (New York: Penguin Press, 2008), pp. 140–43 and *passim*; Dallek, *An Unfinished Life*, pp. 292–93; Branch, *Parting the Waters*, pp. 354–55, 359–71, 374–78, 917–19.

94   Salisbury quoted in Harvard Sitkoff, *The Struggle for Black Equality* (New York: Hill and Wang, 1981), p. 129.

95   On Birmingham and the events of the April and May 1963, see: Victor S. Navasky, *Kennedy Justice* (New York: Atheneum, 1971); Branch, *Parting the Waters*, pp. 586–87, 864–70, chapter 19–21, and *passim*; John Dittmer, *Local People: The Struggle for Civil Rights in Mississippi* (Champaign: University of Illinois Press, 1995), pp. 92–95, 153–57, 165–69; Mackenzie and Weisbrot, *Liberal Hour*, pp. 143–58; Sitkoff, *The Struggle for Black Equality*, pp. 100, 105–8, 111–15, 128–58 and *passim*; Dallek, *An Unfinished Life*, pp. 594–606.

95   "an open test of wills with the Kennedy administration": Victor S. Navasky, *Kennedy Justice*; Branch, *Parting the Waters*, pp. 821–22.

95   "We are confronted primarily": John F. Kennedy, "Radio and Television Report to the American People of Civil Rights," June 11, 1963, http://www.jfklibrary.org/Historical+Resources/Archives/Reference+Desk/Speeches/JFK/003POF03CivilRights06111963.htm.

97 "The very evening of JFK's civil rights speech": Branch, *Parting the Waters*, chapter 22 covers well the events of June–November 1963. Lewis's speech at the March on Washington can be found at http://www.crmvet.org/info/mowjl2.htm; Sitkoff, *Struggle for Black Equality*, p. 165.

98 "You and yours have suffered a great loss": Mrs. A. Marie Lawson to Mrs. John F. Kennedy, no date on letter, envelope postmarked November 23, 1963, Adult Letters, box 9, folder 70, Condolence Mail, JFKL. "if civil rights activists clearly saw": Branch, *Parting the Waters*, p. 880 and chapter 22, *passim*. "I am one person": Carolyn Richards to Mrs. John F. Kennedy, Adult Letters, box 9, folder 65, Condolence Mail, JFKL.

108 "Among the tactics used": J. Morgan Kousser, *The Shaping of Southern Politics: Suffrage Restriction and the Establishment of a One-Party South 1880–1910* (New Haven: Yale University Press, 1974); Alexander Keyssar, *The Right to Vote: The Contested History of Democracy in the United States* (New York: Basic Books, 2000), pp. 111, 130, 228–29, 236–37, 264, 269–71.

141 "Kennedy had proposed to Congress": John F. Kennedy, "Letter to the President of the Senate and the Speaker of the House on Revision of the Immigration Laws," July 23, 1963, *The Public Papers of the Presidents*, http://www.presidency.ucsb.edu/ws/index.php?pid=9355&st=&st1=.

141 "Kennedy published an essay on immigration": John F. Kennedy, "A Nation of Immigrants," *New York Times Magazine*, August 4, 1963, p. 56. Several historians have emphasized the limitations of Kennedy's vision as well as the 1965 immigration reform legislation, which continued discriminatory practices. See, for example, Mae M. Ngai, *Impossible Subjects: Illegal Aliens and the Making of Modern America* (Princeton: Princeton University Press, 2005), esp. chapter 7; Rogers Daniels, *Coming to America: A History of Immigration and Ethnicity in American Life* (New York: HarperPerennial, 2002), chapter 13.

148 "My family had one of the great fortunes": Kennedy quoted in Dallek, *An Unfinished Life*, pp. 30–31.

148 "Kennedy advanced increases": Mackenzie and Weisbrot, *The Liberal Hour*, pp. 87–91 and chapter 3, *passim*.

170 "as he campaigned in West Virginia": Theodore H. White, *The Making of a President 1960* (New York: Atheneum, 1961), pp. 108–25.

178 "Eight inches of fresh snow": "Presidential Inaugural Weather," National Weather Service, Forecast Office Baltimore/Washington, http://www.erh.noaa.gov/lwx/Historic_Events/Inauguration/Inauguration.html#Present-to-Past. "Rejoicing in his youth": "Editorial Comment Across the Nation on President Kennedy's Inauguration," *New York Times*, January 21, 1963, p. 10.

178 "It is the legacy of great men": Dick Santoro to Barbara Longsworth, November 22, 1963, enclosed in Barbara Longsworth to Mrs. John F. Kennedy, January 28, 1964, Personal Remembrances, box 2, folder 13, Condolence Mail, JFKL.

Grief and Loss

202 "The burden of his death": Janet Ott to Mrs. John F. Kennedy , envelope dated November 28, 1963, Adult Letters, box 11, folder 85, Condolence Mail, JFKL. "I am a stranger": Phil Campbell to Mrs. John F. Kennedy, November 28, 1963, Adult Letters, box 10, folder 77, Condolence Mail, JFKL. "The fact of the matter": Alva Adams Thomas to Mrs. John F. Kennedy, December 2, 1963, Adult Letters, box 7, folder 49, Condolence Mail, JFKL. "At first I thought": Blanche McCoy to Mrs. John F. Kennedy, Adult Letters, box 11, folder 83, Condolence Mail, JFKL.

202 "I feel his loss": Mrs. B. Abeur, January 2, 1963 [1964], Adult Letters, box 8, folder 61, Condolence Mail, JFKL. "Our lives and way of living": Mrs. Emma Kelch to Mrs. John F. Kennedy, April 27, 1964, Adult Letters, box 5, folder 39, Condolence Mail, JFKL.

203 "It is terribly difficult for this American": Mrs. Lowell Barry Jacobs to Mrs. John F. Kennedy, January 21, 1964, Adult Letters, box 8, folder 57, Condolence Mail, JFKL.

203 "Maybe I didn't do the proper thing": Mrs. Molesta Lindsay to Mrs. John F. Kennedy, December 10, 1963, Adult Letters, box 16, folder 128, Condolence Mail, JFKL.

206 J. D. Tippit: Manchester, *Death of a President*, pp. 278–82.

223 "the PT boat he commanded": Stephen Plotkin, "Sixty Years Later, the Story of PT 109 Still Captivates," *Prologue*, vol. 35, no. 2 (Summer 2003); Nigel Hamilton, *JFK: Reckless Youth* (New York: Random House, 1992), parts 11–13; Robert J. Donovan, *PT 109: John F. Kennedy in World War II* (New York: McGraw-Hill, 1961). "helped JFK considerably": Dallek, *An Unfinished Life*, pp. 129–30.

223 "Kennedy had no love for": Dallek, *An Unfinished Life*, pp. 89–100, 129–31.

258 "Vaughn Meader was a comedian": "Vaughn Meader, Satirist of Kennedy Family, Dies," *Washington Post*, November 1, 2004, p. B07; John F. Kennedy, News Conference, December 12, 1962, http://www.presidency.ucsb.edu/ws/index.php?pid=9054&st=meader&st1=. "Meader is Dropping Kennedy Imitation," *New York Times*, November 30, 1963, p. 17.

260 "old soldier down in the West Virginia hills": Sgt. Walt Carter, November 27, 1963: Adult Letters, box 7, folder 49.

268 "It is said you are not grown up": Cindy Corwin to Mrs. John F. Kennedy, no date, Adult Letters, box 11, folder 84, Condolence Mail, JFKL.

282 "We shall miss you": Mr. and Mrs. Aldo Angelino to Mrs. John F. Kennedy, November 26, 1963, Adult Letters, box 18, folder 136, Condolence Mail, JFKL.

Photograph Permissions

Politics, Society, and President, 1963

90   Helen Stone to Mrs. John F. Kennedy, December 3, 1963, Adult Letters, box 14, folder 107, Condolence Mail, John F. Kennedy Library. Reprinted with permission of Joyce C. Orman.

96   President Kennedy with leaders of the March on Washington, August 28, 1963, photograph by Cecil Stoughton, John F. Kennedy Library.

121   Photograph of President Kennedy at rally in Allentown, Pennsylvania. Reprinted with permission of Stacey Ryan.

150   Mr. and Mrs. Ralph Cohee to Mrs. John F. Kennedy, December 5, 1963, Adult Letters, box 12, folder 90, John F. Kennedy Library. Reprinted with permission of Shirley Epps.

171   Senator Kennedy shakes hands with a miner, photograph by Hank Walker, with permission of Time-Life Pictures.

179   President Kennedy delivers his inaugural address, January 20, 1961, United States Army Signal Corps Photograph, John F. Kennedy Library.

Grief and Loss

200   R. R. Louk to Mrs. John F. Kennedy, February 14, 1964, Adult Letters, box 20, folder 161, Condolence Mail, John F. Kennedy Library. Reprinted with permission of Linda Ross.

225   Photograph of Charles Harris and John F. Kennedy. Reprinted with permission of Donna Hardy.

243   Monroe Young, Jr., to Mrs. John F. Kennedy, December 1, 1963, Children's Letters, box 63, folder 47, Condolence Mail, John F. Kennedy Library. Reprinted with permission of Lillian Rusk.

269   Lisa Blumberg to Mrs. John F. Kennedy, no date, Adult Letters, box 14, folder 104, Condolence Mail, John F. Kennedy Library. Reprinted with permission of Lisa Blumberg.

272   Douglas K. Grumblatt to Mrs. John F. Kennedy, postmarked November 25, 1963, Children's Letters, box 61, folder 38, Condolence Mail, John F. Kennedy Library. Reprinted with permission of Douglas K. Grumblatt.